Praise for

Time Reborn

"I suspect we're getting a peek here at what the cutting edge of physics a few decades from now is going to look like." *The Boston Globe*

"He challenges not only Einstein's relativity, but also the very notion of natural laws as immutable truths." *The Economist*

"A cornucopia of big ideas, expressed clearly and compassionately. . . . A really rare book: one that is accessible and designed as much for lay readers as it is for physicists, but also a book that proposes new, groundbreaking, theories. . . . Smolin writes like a dream and his tone is spot on for someone who is trying to change a lot of minds about some really big issues." *Santa Barbara Independent*

"Stunning. . . . Applying his deep mastery of cosmology, quantum mechanics, general relativity, and all the diverse attempts at quantum gravity, Lee Smolin weaves a convincing and entirely new view of reality." Stuart Kauffman, University of Vermont, author of *At Home in the Universe*

"[Smolin] is to be applauded for the boldness of his claims. He will no doubt be attacked for presuming to whip away the carpet from beneath the feet of the great and the good of the physics establishment. But it is incredibly refreshing to hear an idea that is quite as original as Smolin's." *Prospect*

"An essay on contemporary debates in important areas of physics, *Time Reborn* is up-to-date and chock-full of diverse hypothetical models. . . . Fascinating." *Winnipeg Free Press*

"Provocative and stimulating new look at the physics of time. . . . Whatever its value as physics—and his ideas may yet prove revolutionary— it is engaging philosophy. . . . The future is unknowable is the point of this thoughtful book, and that's a good thing." Christopher Potter, *The Sunday Times*

ALSO BY LEE SMOLIN

Time Reborn

From the Crisis in Physics
to the Future of the Universe

LEE SMOLIN

VINTAGE CANADA

VINTAGE CANADA EDITION, 2013

Copyright © 2013 Spin Networks, Ltd.

Published in Canada by Vintage Canada, a division of Random House of Canada Limited, Toronto, in 2013. Originally published in hardcover in Canada by Alfred A. Knopf Canada, a division of Random House of Canada Limited, in 2013, and simultaneously in the United States by Houghton Mifflin Harcourt Publishing Company, New York. Distributed by Random House of Canada Limited.

Vintage Canada with colophon is a registered trademark.

www.randomhouse.ca

FIGURES BY HENRY REICH, with the following exceptions: Figure 2, Alamy/ World History Archive. Figure 14 by Carlo Rovelli, used with permission. Figure 21 by R. Loll, J. Ambjørn, and K. N. Anagnostopoulos, used with permission. Figures 13, 15, 16, and 20 by the author.

Library and Archives Canada Cataloguing in Publication

Smolin, Lee, 1955–
      Time reborn : from the crisis in physics to the future of the universe / Lee Smolin.

Originally published: Toronto : Alfred A. Knopf, 2013.
Includes bibliographical references and index.

ISBN 978-0-307-40072-7

      1. Time. I. Title.

BD638.S66 2013   115   C2012-905619-7

Book design by Brian Moore
Cover design by Martha Kennedy
Image credits: (front) © Oleg Moiseyenko/Getty Images; (back) © Stocktrek Images/Getty Images

Printed and bound in the United States of America

10 9 8 7 6 5 4 3 2 1

For my parents, Pauline and Michael

With many thanks to Roberto Mangabeira Unger
for a shared journey

CONTENTS

All things originate from one another,
and vanish into one another
according to necessity . . .
in conformity with the order of time.

—ANAXIMANDER, *On Nature*

PREFACE

WHAT IS TIME?
This deceptively simple question is the single most important problem facing science as we probe more deeply into the fundamentals of the universe. All of the mysteries physicists and cosmologists face — from the Big Bang to the future of the universe, from the puzzles of quantum physics to the unification of the forces and particles — come down to the nature of time.

The progress of science has been marked by the dismissal of illusions. Matter appears to be smooth but turns out to be made of atoms. Atoms seem indivisible but turn out to be built of protons, neutrons, and electrons, the first two of which are made of still more elementary particles called quarks. The sun appears to go around the Earth, but it's the other way around — and when you get right down to it, it turns out that everything moves relative to everything else.

Time is the most pervasive aspect of our everyday experience. Everything we think, feel, or do reminds us of its existence. We perceive the world as a flow of moments that make up our life. But physicists and philosophers alike have long told us (and many people think) that time is the ultimate illusion.

When I ask my nonscientific friends what they think time is, they often answer that its passage is deceptive and whatever is actually real — truth, justice, the divine, scientific laws — lies outside it. The idea that time is an illusion is a philosophical and religious commonplace. For millennia, people have reconciled themselves to life's hardships and our mortality by believing in the possibility of an eventual escape to a timeless and more real world.

Some of our most illustrious thinkers assert the unreality of time. Plato, the greatest philosopher of the ancient world, and Einstein, the greatest physicist of the modern world, both taught a view of nature in which the real is timeless. They saw our experience of time as an accident of our circumstance as human beings — an accident that hides the truth from us. Both believed that the illusion of time must be transcended to perceive the real and the true.

I used to believe in the essential unreality of time. Indeed, I went into physics because as an adolescent I yearned to exchange the time-bound, human world, which I saw as ugly and inhospitable, for a world of pure, timeless truth. Later in life, I discovered that it was pretty nice to be human and the need for transcendent escape faded.

More to the point, I no longer believe that time is unreal. In fact, I have swung to the opposite view: Not only is time real, but nothing we know or experience gets closer to the heart of nature than the reality of time.

My reasons for this *volte-face* lie in science — and, in particular, in contemporary developments in physics and cosmology. I've come to believe that time is the key to the meaning of quantum theory and its eventual unification with space, time, gravity, and cosmology. Most important, I believe that to make sense of the picture of the universe that cosmological observations are bringing to us, we must embrace the reality of time in a new way. This is what I mean by the rebirth of time.

Much of this book sets out the scientific argument for believing in the reality of time. If you are one of the many who believe that time is an illusion, I aim to change your mind. If you already believe that time is real, I hope to give you better reasons for your belief.

This is a book for everyone, because there is no one whose thinking about the world is not shaped by how they see time. Even if you have never pondered its meaning, your thinking — the very language with which you express your thoughts — is colored by ancient metaphysical ideas about time.

When we adopt the revolutionary view that time is real, how we think about everything else will change. In particular, we will tend to see the future in a new way, one that vividly highlights both the opportunities and the dangers confronting the human species.

A small part of the story of this book is the personal journey that led me to rediscover time. My initial motivation might best be described in the language not of science but of fatherhood, through the conversations I have had with my young son, especially when I put him to bed at the end of the day. "Daddy," he asked once as I read to him, "did you have my name when you were my age?" Here was a child awakening to the knowledge that there was a time before him and seeking to connect the short story of his life so far to a longer epic.

Every journey has a lesson to teach, and mine has been to realize just how radical an idea is contained in the simple statement that time is real. Having begun my life in science searching for the equation beyond time, I now believe that the deepest secret of the universe is that its essence rests in how it unfolds moment by moment in time.

❖

There's a paradox inherent in how we think about time. We perceive ourselves as living in time, yet we often imagine that the better aspects of our world and ourselves transcend it. What makes something really true, we believe, is not that it is true now but that it always was and always will be true. What makes a principle of morality absolute is that it holds in every time and every circumstance. We seem to have an ingrained idea that if something is valuable, it exists outside time. We yearn for "eternal love." We speak of "truth" and "justice" as timeless. Whatever we most admire and look up to — God, the truths of mathematics, the laws of nature — is endowed with an existence that transcends time. We act inside time but judge our actions by timeless standards.

As a result of this paradox, we live in a state of alienation from what we most value. This alienation affects every one of our aspirations. In science, experiments and their analysis are time-bound, as are all our observations of nature, yet we imagine that we uncover evidence for timeless natural laws. The paradox also affects our actions as individuals, family members, and citizens, because how we understand time determines how we think about the future.

In this book, I hope to resolve in a new way the paradox of living in time and believing in the timeless. I will propose that time and its passage are fundamental and real and the hopes and beliefs about timeless truths and timeless realms are mythology.

Embracing time means believing that reality consists only of what's real in each moment of time. This is a radical idea, for it denies any kind of timeless existence or truth — whether in the realm of science, morality, mathematics, or government. All those must be reconceptualized, to frame their truths within time.

Embracing time also means that our basic assumptions about how the universe works at the most fundamental level are incomplete. When, in the pages that follow, I assert that time is real, what I'm saying is that:

- Whatever is real in our universe is real in a moment of time, which is one of a succession of moments.
- The past was real but is no longer real. We can, however, interpret and analyze the past, because we find evidence of past processes in the present.
- The future does not yet exist and is therefore open. We can reasonably infer some predictions, but we cannot predict the future completely. Indeed, the future can produce phenomena that are genuinely novel, in the sense that no knowledge of the past could have anticipated them.
- Nothing transcends time, not even the laws of nature. Laws are not timeless. Like everything else, they are features of the present, and they can evolve over time.

In the course of this book, we will see that these hypotheses point to a new direction for fundamental physics — one that I argue is the only way out of the present conundrums of theoretical physics and cosmology. They also have implications for how we should understand our own lives and deal with the challenges humankind faces.

To explain why the reality of time is so consequential, both for science and for matters beyond science, I like to contrast thinking in time with thinking outside time. The idea that truth is timeless and somehow outside the universe is so pervasive that the Brazilian philosopher Roberto Mangabeira Unger refers to it as "the perennial philosophy." It was the essence of Plato's thought, exemplified in the parable, in *Meno,* of the slave boy and the geometry of a square, in which Socrates argues that all discovery is merely recollection.

We think outside time when we imagine that the answer to whatever question we're pondering is out there in some eternal domain of timeless truth. Whether the issue is how to be a better parent or spouse or citizen, or what the optimal organization of society might be, we believe there's something unalterably true out there for us to discover.

Scientists think in time when we conceive of our task as the invention of novel ideas to describe newly discovered phenomena, and of novel mathematical structures to express them. If we think outside time, we believe these ideas somehow existed before we invented them. If we think in time, we see no reason to presume that.

The contrast between thinking in time and outside time is apparent in many arenas of human thought and action. We are thinking outside time when, faced with a technological or social problem, we assume that the possible approaches are already determined, as a set of absolute, pre-existing categories. Anyone who thinks that the correct theory of economics or politics was written down in the century before last is thinking outside time. When we instead see the aim of politics as the invention of novel solutions to novel problems that arise as society evolves, we are thinking in time. We're also thinking in time when we understand that progress in technology, society, and science consists in inventing genuinely new ideas, strategies, and forms of social organization — and trust our ability to do so.

When we unquestioningly accept the strictures, habits, and bureaucracies of our various communities and organizations as if they had an absolute reason to be there, we're trapped outside time. We reenter time when we realize that every feature of a human organization is a result of a history, so that everything about them is negotiable and subject to improvement by the invention of new ways of doing things.

If we believe that the task of physics is the discovery of a timeless mathematical equation that captures every aspect of the universe, then we believe that the truth about the universe lies outside the universe. This is such a familiar habit of thought that we fail to see its absurdity: If the universe is all that exists, then how can something exist outside it for it to be described by? But if we take the reality of time as evident, then there can be no mathematical equation that perfectly captures every aspect of the world, because one property of the real world not shared by any mathematical equation is that it is always some moment.

Darwinian evolutionary biology is the prototype for thinking in time, because at its heart is the realization that natural processes developing in time can lead to the creation of genuinely novel structures. Even novel laws can emerge, when the structures to which they apply come into existence. The principles of sexual selection, for example, could not have come to exist before there were sexes. Evolutionary dynamics has no need of vast abstract spaces, like all the possible viable animals, DNA sequences, sets of proteins, or biological laws. Better, as the theoretical biologist Stuart A. Kauffman proposes, to think of evolutionary dynamics as the exploration in time by the biosphere of what can happen next: the "adjacent possible." The same goes for the evolution of technologies, economies, and societies.

Thinking in time is not relativism but a form of *relationalism* — a philosophy that asserts that the truest description of something consists of specifying its relationships to the other parts of the system it is part of. Truth can be both time-bound and objective when it's about objects that exist once they've been invented, either by evolution or human thought.

On a personal level, to think in time is to accept the uncertainty of life as the necessary price of being alive. To rebel against the precariousness of life, to reject uncertainty, to adopt a zero tolerance to risk, to imagine that life can be organized to completely eliminate danger, is to think outside time. To be human is to live suspended between danger and opportunity.

We try our best to thrive in an uncertain world, to take care of whom and what we love and now and then enjoy ourselves in the process. We make plans, but we can never anticipate fully either the dangers or the opportunities ahead. The Buddhists say that we live in a house we haven't yet noticed is on fire. Danger might arise at any time, and in hunter-gatherer societies it was ever present, but in modern life we have organized things so that it's comparatively rare. The challenge of life is to choose wisely, from the enormous number of possible dangers, what's worth worrying about. It is also about choosing, from all the opportunities that each moment brings, what to do next. We choose where to devote our energy and attention — always in the face of incomplete knowledge of the consequences.

Could we do better? Could we overcome the capriciousness of life and achieve a state wherein we knew, if not everything, enough to see all the consequences of our choices — the dangers and the opportunities alike? That is, could we live a truly rational life, without surprises? If time were an illusion, we could imagine this as possible, because in a world in which time was dispensable there would be no fundamental difference between knowledge of the present and knowledge of the future. It would take just a bit more computation to work out. Some number, some formula, could be computed and decoded to tell us all we needed to know.

But if time is real, the future is not determinable from knowledge of the present. There is no escape from our situation, no redemption from the surprises that come from living in ignorance of most of the consequences of our actions. Surprise is inherent in the structure of the world. Nature can throw us surprises for which no amount of knowledge would have prepared us. Novelty is real. We can create, with our imagination, outcomes not computable from knowledge of the present. This is why it matters for each of us whether time is real or not: The answer can change how we view our situation as seekers of happiness and meaning in a largely unknown universe. I will return to these themes in the Epilogue, where I suggest that the reality of time can help us think about such challenges as climate change and economic crisis.

Before we begin the main argument of the book, a few words of advice.

I have tried to make the arguments of this book accessible for the general reader without a background in physics or mathematics. There are no equations, and everything you need to know to follow my arguments is explained. The essential questions are illustrated with the simplest examples possible. As we move on to more sophisticated subjects, readers are advised, if confused, to do what scientists learn to do, which is to skim or skip ahead to a point where the text becomes clearer to them. Readers wanting more background can also consult the several appendices, which are available on-line at www.time reborn.com. The reader may also find it helpful to consult the Notes, which contain citations, helpful remarks either for laypeople or experts, and further discussions that may interest some readers.

My own journey back to time has taken more than twenty years, from my recognition that laws are to be explained by their having evolved, through my struggles with relativity, quantum foundations, and quantum gravity, which finally led me to the view described here. Collaborations and conversations with several friends and colleagues have been essential to my progress on this road; they are detailed in the Acknowledgments and Notes, as is my use of the results and ideas of others. None of these interactions was more important than a fruitful and provocative collaboration with Roberto Mangabeira Unger, during which we formulated the main argument and many of the key ideas that follow.[1]

Readers should be aware that there are many points of view about time, quantum theory, cosmology, and other such topics that are not discussed here. There is a vast literature by physicists, cosmologists, and philosophers concerning the issues I touch on. This does not pretend to be an academic book. I have chosen to give readers who may be encountering this area of discussion for the first time one path through its complex terrain, highlighting particular arguments that are its focus.[2] There are (to take one example) bookshelves full of writings analyzing Kant's views on space and time, which are not mentioned here. Nor do I describe some of the views of contemporary philosophers. I ask forgiveness of my learned friends for these omissions and direct the interested reader to the Bibliography, which contains suggestions for further reading about time.

LEE SMOLIN
TORONTO, AUGUST 2012

INTRODUCTION

THE SCIENTIFIC CASE for time being an illusion is formidable. That is why the consequences of adopting the view that time is real are revolutionary.

The core of the physicists' case against time relies on the way we understand what a law of physics is. According to this dominant view, everything that happens in the universe is determined by a law, which dictates precisely how the future evolves out of the present. The law is absolute and, once present conditions are specified, there is no freedom or uncertainty in how the future will evolve.

As Thomasina, the precocious heroine of Tom Stoppard's play *Arcadia*, explains to her tutor: "If you could stop every atom in its position and direction, and if your mind could comprehend all the actions thus suspended, then if you were really, really good at algebra you could write the formula for all the future; and although nobody can be so clever as to do it, the formula must exist just as if one could."

I used to believe that my job as a theoretical physicist was to find that formula; I now see my faith in its existence as more mysticism than science.

Were he writing lines for a modern character, Stoppard would have had Thomasina say that the universe is like a computer. The laws of physics are the program. When you give it an input — the present positions of all the elementary particles in the universe — the computer runs for an appropriate amount of time and gives you the output, which is all the positions of the elementary particles at some future time. Within this view of nature, nothing happens except the rearrangement of particles according to timeless laws, so according to these laws the future is already completely determined by the present, as the present was by the past.

This view diminishes time in several ways.[1] There can be no surprises, no truly novel phenomena, because all that happens is rearrangement of the atoms. The properties of the atoms themselves are timeless, as are the laws controlling them; neither ever changes. Any feature of the world at a future time can be computed from the configuration of the present. That is, the passage of time can be replaced by a computation, which means that the future is logically a consequence of the present.

Einstein's theories of relativity make even stronger arguments that time is inessential to a fundamental description of the world, as I'll discuss in chapter 6. Relativity strongly suggests that the whole history of the world is a timeless unity; present, past, and future have no meaning apart from human subjectivity. Time is just another dimension of space, and the sense we have of experiencing moments passing is an illusion behind which is a timeless reality.

These assertions may seem horrifying to anyone whose worldview includes a place for free will or human agency. This is not an argument I will engage in here; my case for the reality of time rests purely on science. My job will be to explain why the usual arguments for a predetermined future are wrong scientifically.

In Part I, I will present the case from science for believing that time is an illusion. In Part II, I will demolish those arguments and show why time must be taken to be real if fundamental physics and cosmology are to overcome the crises they currently face.

To frame the argument of Part I, I trace the development of the concepts of time used in physics, from Aristotle and Ptolemy through Galileo, Newton, Einstein, and on to our contemporary quantum cosmologists, and show how our concept of time was diminished, step by step, as physics progressed. Telling the story this way also allows me to gently introduce the material the lay reader needs for an understanding of the argument. Indeed, key points can be introduced by ordinary examples of balls falling and planets orbiting. Part II tells a more contemporary story, since the argument that time must be reinserted into the core of science arose as a result of recent developments.

My argument starts with a simple observation: The success of scientific theories from Newton through the present day is based on their use of a particular framework of explanation invented by Newton. This framework views nature as consisting of nothing but particles with timeless properties, whose motions and interactions are determined by timeless laws. The properties of the particles, such as their masses and electric charges, never change, and neither do the laws that act on them. This framework is ideally suited to describe small parts of the universe, but *it falls apart when we attempt to apply it to the universe as a whole.*

All the major theories of physics are about parts of the universe — a radio, a ball in flight, a biological cell, the Earth, a galaxy. When we describe a part of the universe, we leave ourselves and our measuring tools outside the system. We leave out our role in selecting or preparing the system we study. We leave out the references that serve to establish where the system is. Most crucially for our concern with the nature of time, we leave out the clocks by which we measure change in the system.

The attempt to extend physics to cosmology brings new challenges that require fresh thinking. A cosmological theory cannot leave anything out. To be complete, it must take into account everything in the universe, including ourselves as observers. It must account for our measuring instruments and clocks. When we do cosmology, we confront a novel circumstance: It is impossible to get outside the system we're studying when that system is the entire universe.

Moreover, a cosmological theory must do without two important aspects of the methodology of science. A basic rule of science is that an experiment must be done many times to be sure of the result. But we cannot do this with the universe as a whole — the universe only happens once. Nor can we prepare the system in different ways and study the consequences. These are very real handicaps, which make it much harder to do science at the level of the universe as a whole.

Nonetheless, we want to extend physics to a science of cosmology. Our first instinct is to take the theories that worked so well when applied to small parts of the universe and scale them up to describe the universe as a whole. As I'll show in chapters 8 and 9, this cannot work. The Newtonian framework of timeless laws acting on particles with timeless properties is unsuited to the task of describing the entire universe.

Indeed, as I will show in detail, the very features that make these kinds of theories so successful when applied to small parts of the universe cause them to fail when we attempt to apply them to the universe as a whole.

I realize that this assertion goes counter to the practice and hopes of many colleagues, but I ask only that the reader pay close attention to the case I make for it in Part II. There I will show in general, and illustrate by specific example, that when we attempt to scale up our standard theories to a cosmological theory, we are rewarded with dilemmas, paradoxes, and unanswerable questions. Among these are the failure of any standard theory to account for the choices made in the early universe — choices of initial conditions and choices of the laws of nature themselves.

Some of the literature of contemporary cosmology consists of the efforts of very smart people to wrestle with these dilemmas, paradoxes, and unanswerable questions. The notion that our universe is part of a vast or infinite multiverse is popular — and understandably so, because it is based on a methodological error that is easy to fall into. Our current theories can work at the level of the universe only if our universe is a subsystem of a larger system. So we invent a fictional environment and fill it with other universes. This cannot lead to any real scientific progress, because we cannot confirm or falsify any hypothesis about universes causally disconnected from our own.[2]

The purpose of this book is to suggest that there is another way. We need to make a clean break and embark on a search for a new kind of theory that can be applied to the whole universe — a theory that avoids the confusions and paradoxes, answers the unanswerable questions, and generates genuine physical predictions for cosmological observations.

I do not have such a theory, but what I can offer is a set of principles to guide the search for it. These are presented in chapter 10. In the chapters that follow it, I will illustrate how the principles can inspire new hypotheses and models of the universe that point the way to a true cosmological theory. The central principle is that time must be real and physical laws must evolve in that real time.

The idea of evolving laws is not new, nor is the idea that a cosmological science will require them.[3] The American philosopher Charles Sanders Peirce wrote in 1891:

> To suppose universal laws of nature capable of being apprehended by the mind and yet having no reason for their special forms, but standing inexplicable and irrational, is hardly a justifiable position. Uniformities are precisely the sort of facts that need to be accounted for. . . . Law is par excellence the thing that wants a reason.
>   Now the only possible way of accounting for the laws of nature and for uniformity in general is to suppose them results of evolution."[4]

The contemporary philosopher Roberto Mangabeira Unger has more recently proclaimed:

> You can trace the properties of the present universe back to properties it must have had at its beginning. But you cannot show that these are the only properties that any universe might have had. . . . Earlier or later universes might have had entirely different laws. . . . To state the laws of nature is not to describe or to explain all possible histories of all possible universes. Only a relative distinction exists between lawlike explanation and the narration of a one-time historical sequence."[5]

Paul Dirac, who ranks with Einstein and Niels Bohr as one of the most consequential physicists of the 20th century, speculated: "At the beginning of time the laws of Nature were probably very different from what they are now. Thus, we should consider the laws of Nature as continually changing with the epoch, instead of as holding uniformly

throughout space-time."[6] John Archibald Wheeler, one of the great
American physicists, also imagined that laws evolved. He proposed
that the Big Bang was one of a series of events within which the laws of
physics were reprocessed. He also wrote, "There is no law except the
law that there is no law."[7] Even Richard Feynman, another of the great
American physicists and Wheeler's student, once mused in an inter-
view: "The only field which has not admitted any evolutionary question is physics. Here are the laws, we say, . . . but how did they get that
way, in time? . . . So, it might turn out that they are not the same [laws]
all the time and that there is a historical, evolutionary, question."[8]

In my 1997 book, *The Life of the Cosmos,* I proposed a mechanism
for laws to evolve, which I modeled on biological evolution.[9] I imagined that universes could reproduce by forming baby universes inside black holes, and I posited that whenever this happens, the laws of
physics change slightly. In this theory, the laws played the role of genes
in biology; a universe was seen as an expression of a choice of laws
made at its formation, just as an organism is an expression of its genes.
Like the genes, the laws could mutate randomly from generation to
generation. Inspired by then-recent results of string theory, I imagined that the search for a fundamental unified theory would lead not
to a single Theory of Everything but to a vast space of possible laws. I
called this the landscape of theories, taking the language from population genetics, whose practitioners work with fitness landscapes. I will
not say more about this here, as it is the subject of chapter 11, except to
say that this theory, cosmological natural selection, makes several predictions that, remarkably, have held up despite several opportunities
to falsify them in the years since.

Over the last decade, many string theorists have embraced the concept of a landscape of theories. As a result, the question of how the
universe chooses which laws to follow has become especially urgent.
This, I will argue, is one of the questions that can be answered only
within a new framework for cosmology in which time is real and laws
evolve.

Laws, then, are not imposed on the universe from outside it. No
external entity, whether divine or mathematical, specifies in advance

what the laws of nature are to be. Nor do the laws of nature wait, mute, outside of time for the universe to begin. Rather the laws of nature emerge from inside the universe and evolve in time with the universe they describe. It is even possible that, just as in biology, novel laws of physics may arise as regularities of new phenomena that emerge during the universe's history.

Some might see the disavowal of eternal laws as a retreat from the goals of science. But I see it as the jettisoning of excess metaphysical baggage that weighs down our search for truth. In the coming chapters, I will provide examples illustrating how the idea of laws evolving in time leads to a more scientific cosmology — by which I mean one more generative of predictions subject to experimental test.

◆

To my knowledge, the first scientist since the dawn of the Scientific Revolution to think really hard about how to make a theory of a whole universe was Gottfried Wilhelm Leibniz, who, among other things, was Newton's rival, famously in the matter of which of them was the first to invent the calculus. He also anticipated modern logic, developed a system of binary numbers, and much else. He has been called the smartest person who ever lived. Leibniz formulated a principle to frame cosmological theories called the *principle of sufficient reason,* which states that there must be a rational reason for every apparent choice made in the construction of the universe. Every query of the form, "Why is the universe like X rather than Y?" must have an answer. So if a God made the world, He could not have had any choice in the blueprint. Leibniz's principle has had a profound effect on the development of physics so far, and, as we will see, it continues to be reliable as a guide in our efforts to devise a cosmological theory.

Leibniz had a vision of a world in which everything lives not in space but immersed in a network of relationships. These relationships define space, not the reverse. Today the idea of a universe of connected, networked entities pervades modern physics, as well as biology and computer science.

In a relational world (which is what we call a world where relationships precede space), there are no spaces without things. Newton's concept of space was the opposite, for he understood space to be absolute. This means atoms are defined by where they are in space but space is in no way affected by the motion of atoms. In a relational world, there are no such asymmetries. Things are defined by their relationships. Individuals exist, and they may be partly autonomous, but their possibilities are determined by the network of relationships. Individuals encounter and perceive one another through the links that connect them within the network, and the networks are dynamic and ever evolving.

As I will explain in chapter 3, it follows from Leibniz's great principle that there can be no absolute time that ticks on blindly whatever happens in the world. Time must be a consequence of change; without alteration in the world, there can be no time. Philosophers say that time is relational — it is an aspect of relations, such as causality, that govern change. Similarly, space must be relational; indeed, every property of an object in nature must be a reflection of dynamical[10] relations between it and other things in the world.

Leibniz's principles contradicted the basic ideas of Newtonian physics, so it took some time for them to be fully appreciated by working scientists. It was Einstein who embraced Leibniz's legacy and used his principles as major motivation for his overthrow of Newtonian physics and its replacement by general relativity, a theory of space, time, and gravity that goes far to instantiate Leibniz's relational view of space and time. Leibniz's principles are also realized in a different way in the parallel quantum revolution. I call the 20th-century revolution in physics the relational revolution.

The problem of unifying physics and, in particular, bringing together quantum theory with general relativity into one framework is largely the task of completing the relational revolution in physics. The main message of this book is that this requires embracing the ideas that time is real and laws evolve.

The relational revolution is already in full swing in the rest of science. Darwin's revolution in biology is one front, manifested both in the notion of a species being defined by its relation to all the other organisms in its environment and in the concept that a gene's action is defined only in the context of the network of genes regulating its action. As we are quickly coming to realize, biology is about information, and there is no more relational concept than information, relying as it does on a relationship between the sender and receiver at each end of a communications channel.

In the social sphere, the liberal concept of a world of autonomous individuals (conceived by the philosopher John Locke as analogous to the physics of his friend Isaac Newton) is being challenged by a view of society as composed of interdependent individuals, only partly autonomous, whose lives are meaningful only within a skein of relationships. The new informational halo within which we are so recently enmeshed expresses the relational idea through the metaphor of the network. As social beings, we see ourselves as nodes in a network whose connections define us. Today the idea of a social system made up of connected, networked entities increasingly crops up in social theories formulated by everyone from feminist political philosophers to management gurus. How many users of Facebook are aware that their social lives are now organized by a potent scientific idea?

The relational revolution is already far along. At the same time, it is clearly in crisis. On some fronts, it's stuck. Wherever it is in crisis, we find three kinds of questions under hot debate. What is an individual? How do novel kinds of systems and entities emerge? How are we to usefully understand the universe as a whole?

The key to these puzzles is that neither individuals, systems, nor the universe as a whole can be thought of as things that simply are. They are all compounded by processes that take place in time. The missing element, without which we cannot answer these questions, is to see them as processes developing in time. I will argue that to succeed, the relational revolution must embrace the notion of time and the present moment as a fundamental aspect of reality.

In the old way of thinking, individuals were just the smallest units in a system, and if you wanted to understand how a system worked you took it apart and studied how its parts behaved. But how are we to understand the properties of the most fundamental entities? They have no parts, so reductionism (as this method is called) gets us no further. The atomic viewpoint has no place to go here; it, too, is truly stuck. This is a great opportunity for the nascent relational program, for it can — and indeed must — seek the explanation for properties of elementary particles in the network of their relations.

This is already happening in the unified theories we have so far. In the Standard Model of Particle Physics, which is the best theory we have so far of the elementary particles, the properties of an electron, such as its mass, are dynamically determined by the interactions in which it participates. The most basic property a particle can have is its mass, which determines how much force is needed to change its motion. In the Standard Model, all the particles' masses arise from their interactions with other particles and are determined primarily by one — the Higgs particle. No longer are there absolutely "elementary" particles; everything that behaves like a particle is, to some extent, an emergent consequence of a network of interactions.

Emergence is an important term in a relational world. A property of something made of parts is emergent if it would not make sense when attributed to any of the parts. Rocks are hard, and water flows, but the atoms they're made of are neither solid nor wet. An emergent property will often hold approximately, because it denotes an averaged or high-level description that leaves out much detail.

As science progresses, aspects of nature once considered fundamental are revealed as emergent and approximate. We once thought that solids, liquids, and gases were fundamental states; now we know that these are emergent properties, which can be understood as different ways to arrange the atoms that make up everything. Most of the laws of nature once thought of as fundamental are now understood as emergent and approximate. Temperature is just the average energy of atoms in random motion, so the laws of thermodynamics that refer to temperature are emergent and approximate.

I'm inclined to believe that just about everything we now think is fundamental will also eventually be understood as approximate and emergent: gravity and the laws of Newton and Einstein that govern it, the laws of quantum mechanics, even space itself.

The fundamental physical theory we seek will not be about things moving in space. It will not have gravity or electricity or magnetism as fundamental forces. It will not be quantum mechanics. All these will emerge as approximate notions when our universe grows large enough.

If space is emergent, does that mean that time is also emergent? If we go deep enough into the fundamentals of nature, does time disappear? In the last century, we have progressed to the point where many of my colleagues consider time to be emergent from a more fundamental description of nature in which time does not appear.

I believe — as strongly as one can believe anything in science — that they're wrong. Time will turn out to be the only aspect of our everyday experience that *is* fundamental. The fact that it is always some moment in our perception, and that we experience that moment as one of a flow of moments, is not an illusion. It is the best clue we have to fundamental reality.

PART I

WEIGHT: THE EXPULSION OF TIME

Falling

BEFORE STARTING THIS or any other journey of discovery, we should heed the advice of the Greek philosopher Heraclitus, who, barely a few steps into the epic story that is science, had the wisdom to warn us that "Nature loves to hide." And indeed she does; consider that most of the forces and particles that science now considers fundamental lay hidden within the atom until the last century. Some of Heraclitus's contemporaries spoke of atoms, but without really knowing whether or not they existed. And their concept was wrong, for they imagined atoms as indivisible. It took until Einstein's papers of 1905 for science to catch up and form the consensus that matter is made of atoms. And six years later the atom itself was broken into pieces. Thus began the unraveling of the interior of atoms and the discoveries of the worlds hidden within.

The largest exception to the modesty of nature is gravity. It is the only one of the fundamental forces whose effects everyone observes with no need for special instruments. Our very first experiences of struggle and failure are against gravity. Consequently, gravity must have been among the first natural phenomena to be named by our species.

Nonetheless, key aspects of the common experience of falling remained hidden in plain sight until the dawn of science, and much remains hidden still. As we shall see in later chapters, one thing that remains hidden about gravity is its relation to time. So we start our journey toward the discovery of time with falling.

◆

"Why can't I fly, Daddy?"

We were on the top deck, looking down three floors to the back garden.

"I'll just jump off and fly down to Mommy in the garden, like those birds."

"Bird" had been his first word, uttered at the sparrows fluttering in the tree outside his nursery window. Here is the elemental conflict of parenthood: We want our children to feel free to soar beyond us, but we also fear for their safety in an uncertain world.

I told him sternly that people can't fly and he was absolutely never to try, and he burst into tears. To distract him, I took the opportunity to tell him about gravity. Gravity is what holds us down to Earth. It is why we fall, and why everything else falls.

The next word out of his mouth was, unsurprisingly, "Why?" Even a three-year-old knows that to name a phenomenon is not to explain it.

But we could play a game to see *how* things fall. Soon we were throwing all kinds of toys down into the garden, doing "speriments" to see whether they all fell the same way or not. I quickly found myself thinking of a question that transcends the powers of a three-year-old mind. When we throw an object and it falls as it moves away from us, it traces a curve in space. What sort of curve is it?

It's not surprising that this question doesn't occur to a three-year-old. It doesn't seem to have been an important question for thousands of years after we regarded ourselves as highly civilized. It seems that Plato and some of the other great philosophers of the ancient world were content to watch things fall around them without wondering whether falling bodies travel along a specific kind of curve. The one

ancient philosopher who did speculate about the paths taken by falling bodies — Aristotle — proposed an answer that was easy to disprove
but nonetheless was blindly believed for more than a thousand years.

The first person to understand correctly the paths traced by falling bodies was the Italian Galileo Galilei, early in the 17th century. He
presented his results in *Dialogue Concerning Two New Sciences*, which
he wrote during his seventies, when he was under house arrest by the
Inquisition. In this book, he reported that falling bodies always travel
along the same sort of curve, which is a parabola.

Galileo not only discovered how objects fall but also explained his
discovery. The fact that falling bodies trace parabolas is a direct consequence of another fact he was the first to observe, which is that all
objects, whether thrown or dropped, fall with a constant acceleration.

Galileo's observation that all falling objects trace a parabola is one
of the most wonderful discoveries in all of science. Falling is universal,
and so is the kind of curve that falling bodies trace. It doesn't matter what the object is made of, how it is put together, or what its function is. Nor does it matter how many times, from what height, or with
what forward speed we drop or throw the object. We can repeat the
experiment over and over, and each time it's a parabola. The parabola
is one of the simplest curves to describe. It is the set of points equidistant from a point and a line. So one of the most universal phenomena
is also one of the simplest.

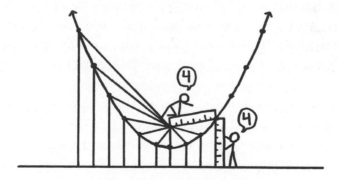

Figure 1: Definition of a parabola: the points equidistant from a point
and a line.

A parabola is a concept from mathematics — an example of what we call a mathematical object — that was known to mathematicians well before Galileo's time. Galileo's observation that bodies fall along parabolas is one of the first examples we have of a law of nature — that is, a regularity in the behavior of some small subsystem of the universe. In this case, the subsystem is an object falling near the surface of a planet. This has happened a great number of times and in a great number of places since the universe began; hence there are many instances to which the law applies.

Here's a question children may ask when they're a bit older: What does it say about the world that falling objects trace such a simple curve? Why should a mathematical concept like a parabola, an invention of pure thought, have anything to do with nature? And why should such a universal phenomenon as falling have a mathematical counterpart that is one of the simplest and most beautiful curves in all of geometry?

◈

Since Galileo's discovery, physicists have profitably used mathematics in the description of physical phenomena. It may seem obvious to us now that a law must be mathematical, but it took almost 2,000 years after Euclid codified his axioms of geometry for someone to propose a correct mathematical law applying to the motion of objects on Earth. From the time of the ancient Greeks to the 17th century, educated people knew what a parabola was, but not a single one of them seems to have wondered whether the balls, arrows, and other objects they dropped, flung, or shot fell along any particular sort of curve.[1] Any one of them could have made Galileo's discovery; the tools he used were available in the Athens of Plato and the Alexandria of Hypatia. But nobody did. What changed to make Galileo think that mathematics had a role in describing something as simple as how things fall?

This question takes us into the heart of some questions easy to state but hard to answer: What is mathematics about? Why does it come into science?

Mathematical objects are constituted out of pure thought. We don't discover parabolas in the world, we invent them. A parabola or a circle or a straight line is an idea. It must be formulated and then captured in a definition. *"A circle is a set of points equidistant from a single point. . . . A parabola is a set of points equidistant from a point and a line."* Once we have the concept, we can reason directly from the definition of a curve to its properties. As we learned in high school geometry class, this reasoning can be formalized in a proof, each argument of which follows from earlier arguments by simple rules of reasoning. At no stage in this formal process of reasoning is there a role for observation or measurement.[2]

A drawing can approximate the properties demonstrated by a proof, but always imperfectly. The same is true of curves we find in the world: the curve of a cat's back when she stretches or the sweep of the cables of a suspension bridge. They will only approximately trace a mathematical curve; when we look closer, there's always some imperfection in the realization. Thus the basic paradox of mathematics: The things it studies are unreal, yet they somehow illuminate reality. But how? The relationship between reality and mathematics is far from evident, even in this simple case.

You may wonder what an exploration of mathematics has to do with an exploration of gravity. But this is a necessary digression, because mathematics is as much at the heart of the mystery of time as gravity is, and we need to sort out how mathematics relates to nature in a simple case, such as bodies falling along curves. Otherwise when we get to the present era and encounter statements like "The universe is a four-dimensional spacetime manifold," we will be rudderless. Without having navigated waters shallow enough for us to see bottom, we'll be easy prey to mystifiers who want to sell us radical metaphysical fantasies in the guise of science.

Although perfect circles and parabolas are never to be found in nature, they share one feature with natural objects: a resistance to manipulation by our fantasy and our will. The number pi — the ratio of a circle's circumference to its diameter — is an idea. But once the concept was invented, its value became an objective property, one that must be discovered by further reasoning. There have been attempts to legislate the value of pi, and they have revealed a profound misunderstanding. No amount of wishing will make the value of pi anything other than it is. The same is true for all the other properties of curves and other objects in mathematics; these objects are what they are, and we can be right or wrong about their properties but we can't change them.

Most of us get over our inability to fly. We eventually concede that we have no influence on many of the aspects of nature. But isn't it a bit unsettling that there are concepts existing only in our minds whose properties are as objective and immune to our will as things in nature? We invent the curves and numbers of mathematics, but once we have invented them we cannot alter them.

But even if curves and numbers resemble objects in the natural world in the stability of their properties and their resistance to our will, they are not the same as natural objects. They lack one basic property shared by every single thing in nature. Here in the real world, it is always some moment of time. Everything we know of in the world participates in the flow of time. Every observation we make of the world can be dated. Each of us, and everything we know of in nature, exists for an interval of time; before and after that interval, we and they do not exist.

Curves and other mathematical objects do not live in time. The value of pi does not come with a date before which it was different or undefined and after which it will change. If it's true that two parallel lines never meet in the plane as defined by Euclid, it always was and always will be true. Statements about mathematical objects like curves and numbers are true in a way that doesn't need any qualification with regard to time. Mathematical objects transcend time. But how can anything exist without existing in time?[3]

People have been arguing about these issues for millennia, and philosophers have yet to reach agreement about them. But one proposal has been on the table ever since these questions were first debated. It holds that curves, numbers, and other mathematical objects exist just as solidly as what we see in nature — except that they are not in our world but in another realm, a realm without time. So there are *not* two kinds of things in our world, time-bound things and timeless things. There are, rather, two worlds: a world bound in time and a timeless world.

The idea that mathematical objects exist in a separate, timeless world is often associated with Plato. He taught that when mathematics speaks of a triangle, it is not any triangle in the world but an ideal triangle, which is just as real (and even more so) but exists in another realm, one outside time. The theorem that the angles of a triangle add up to 180 degrees is not precisely true of any real triangle in our physical world, but it is absolutely and precisely true of that ideal mathematical triangle existing in the mathematical world. So when we prove the theorem, we are gaining knowledge of something that exists outside time and demonstrating a truth that, likewise, is not bounded by present, past, or future.

If Plato is right, then simply by reasoning we human beings can transcend time and learn timeless truths about a timeless realm of existence. Some mathematicians claim to have deduced certain knowledge about the Platonic realm. This claim, if true, gives them a trace of divinity. How do they imagine they pulled this off? Is their claim credible?

When I want a dose of Platonism, I ask my friend Jim Brown for lunch. Both of us enjoy a good meal, during which he will patiently, and not for the first time, explain the case for belief in the timeless reality of the mathematical world. Jim is unusual among philosophers in coupling a razor-sharp mind with a sunny disposition. You sense that he's happy in life, and it makes you happy to know him. He's a good philosopher; he knows all the arguments on each side, and he has no trouble discussing those he can't refute. But I haven't found a way to challenge his confidence in the existence of a timeless realm of mathematical objects. I sometimes wonder if his belief in truths beyond the ken of humans contributes to his happiness at being human.

One question that Jim and other Platonists admit is hard for them to answer is how we human beings, who live bounded in time, in contact only with other things similarly bounded, can have definite knowledge of the timeless realm of mathematics. We get to the truths of mathematics by reasoning, but can we really be sure our reasoning is correct? Indeed, we cannot. Occasionally errors are discovered in the proofs published in textbooks, so it's likely that errors remain. You can try to get out of the difficulty by asserting that mathematical objects don't exist at all, even outside time. But what sense does it make to assert that we have reliable knowledge about a domain of nonexistent objects?

Another friend I discuss Platonism with is the English mathematical physicist Roger Penrose. He holds that the truths of the mathematical world have a reality not captured by any system of axioms. He follows the great logician Kurt Gödel in arguing that we can reason directly to truths about the mathematical realm — truths that are beyond formal axiomatic proof. Once, he said something like the following to me: "You're certainly sure that one plus one equals two. That's a fact about the mathematical world that you can grasp in your intuition and be sure of. So one-plus-one-equals-two is, by itself, evidence enough that reason can transcend time. How about two plus two equals four? You're sure of that, too! Now, how about five plus five equals ten? You have no doubts, do you? So there are a very large number of facts about the timeless realm of mathematics that you're confident you know." Penrose believes that our minds can transcend the ever changing flow of experience and reach a timeless eternal reality behind it.[4]

We discovered the phenomenon of gravity when we realized that our experience of falling is an encounter with a universal natural occurrence. In our attempts to comprehend this phenomenon, we discerned an amazing regularity: All objects fall along a simple curve the ancients invented called parabolas. Thus we can relate a universal phenomenon affecting time-bound things in the world with an invented concept that, in its perfection, suggests the possibility of truths — and of existence — outside time. If you're a Platonist, like Brown and Penrose, the discovery that bodies universally fall along parabolas is no less than the perception of a relationship between our earthly time-bound world and another, timeless world of eternal truth and beauty.

Galileo's simple discovery then takes on a transcendental or religious significance: It is the discovery of a reflection of timeless divinity acting universally in our world. The falling of a body in time in our imperfect world reveals a timeless essence of perfection at nature's heart.

This vision of transcendence to the timeless via science has drawn many into science, including myself, but now I'm sure it's wrong. The dream of transcendence has a fatal flaw at its core, related to its claim to explain the time-bound by the timeless. Because we have no physical access to the imagined timeless world, sooner or later we'll find ourselves just making stuff up (I'll present you with examples of this failing in chapters to come). There's a cheapness at the core of any claim that our universe is ultimately explained by another, more perfect world standing apart from everything we perceive. If we succumb to that claim, we render the boundary between science and mysticism porous.

Our desire for transcendence is at root a religious aspiration. The yearning to be liberated from death and from the pains and limitations of our lives is the fuel of religions and of mysticism. Does the seeking of mathematical knowledge make one a kind of priest, with special access to an extraordinary form of knowledge? Should we simply recognize mathematics for the religious activity it is? Or should we be concerned when the most rational of our thinkers, the mathematicians, speak of what they do as if it were the route to transcendence from the bounds of human life?

It is far more challenging to accept the discipline of having to explain the universe we perceive and experience only in terms of itself—to explain the real only by the real, and the time-bound only by the time-bound. But although it's more challenging, this restricted, less romantic route will ultimately be the more successful. The prize that awaits us is to understand, finally, the meaning of time on its own terms.

2

The Disappearance of Time

GALILEO WAS NOT the first to associate motion with curves. He was just the first to do it for motion on Earth. One reason it may never have occurred to anyone before Galileo that bodies fall on parabolas is that no one had perceived those parabolas directly. The paths of falling bodies were simply too fast to see.[1] But long before Galileo, people did have examples of motion slow enough to easily record. These were the motions of the sun, moon, and planets in the sky. Plato and his students had records of their positions, which the Egyptians and Babylonians had been keeping for thousands of years.

Such records amazed and delighted those who studied them, because they contained patterns — some obvious, like the annual motion of the sun, and others far from obvious, like the cycle of eighteen years and eleven days found in records of solar eclipses. These patterns were clues to the true constitution of the universe the ancients found themselves in. Over many centuries, scholars worked to decipher them, and it is by these efforts that mathematics first entered science.

But this isn't the whole answer. Galileo used no tool not available to the Greeks, so there must have been some conceptual reason for the lack of progress on earthly motion. Did Galileo's predecessors have some blind spot about motion on Earth that Galileo lacked? What did they believe that he didn't?

Let's consider the discovery of one of the simplest and most profound patterns found by ancient astronomers. The word "planet" comes from the Greek word for wanderer, but the planets don't wander all over the sky. They all move along a great circle called the ecliptic, which is fixed with respect to the stars. The discovery of the ecliptic must have been the first step in decoding the records of planetary positions.

A circle is a mathematical object, defined by a simple rule. What does it mean if a circle is seen in the motions in the sky? Is this the visitation of a timeless phenomenon into the ephemeral, time-bound world? This might be how we would see it, but this is not how the ancients understood it. The universe, for the ancients, was split into two realms: the earthly realm, which was the arena of birth and death, of change and decay, and the heavenly realm above, which was a place of timeless perfection. For them the sky was already a transcendental realm; it was populated by divine objects that neither grew nor decayed. This was, after all, what they observed. Aristotle himself noted that "in the whole range of time past, so far as our inherited records reach, no change appears to have taken place either in the whole scheme of the outermost Heaven or in any of its proper parts."[2]

If the objects in this divine realm were to move, these movements could only be perfect and thus eternal. To the ancients, it was evident that the planets move along a circle because, being divine and perfect, they could move only on the curve that was the most perfect. But the earthly realm is not perfect, so it might have seemed bizarre to them to describe motion on Earth in terms of perfect mathematical curves.

The division of the world into an earthly realm and heavenly spheres was codified in Aristotelian physics. Everything in the earthly realm was composed of mixtures of four elements: earth, air, fire, and water. Each had a natural motion: The natural motion of earth, for example, was to seek the center of the universe. Change followed from the mixing of these four essences. Aether was a fifth element, the quintessence, which made up the heavenly realm and the objects that moved across it.

This division was the origin of the connection of elevation with transcendence. God, the heavens, perfection — these are above us, while we are trapped here below. From this perspective, the discovery that mathematical shapes are traced by motions in the sky makes sense, because both the mathematical and the heavenly are realms that transcend time and change. To know each of them is to transcend the earthly realm.

Mathematics, then, entered science as an expression of a belief in the timeless perfection of the heavens. Useful as mathematics has turned out to be, the postulation of timeless mathematical laws is never completely innocent, for it always carries a trace of the metaphysical fantasy of transcendence from our earthly world to one of perfect forms.

Long after science has moved on from the cosmos of the ancients, its basic shape influences everyday speech and metaphor. We speak of rising to the occasion. We look upward for inspiration. Whereas to fall (as in "falling in love," for instance) means to surrender to loss of control. More than that, the opposition of "ascending" and "falling" symbolizes the conflict between the corporeal and the spiritual. Heaven is above us, Hell is below. When we degrade ourselves, we sink downward into the earth. God, and everything we ultimately seek, is above us.

Music was another way the ancients experienced transcendence. Listening to music, we often experience a profound beauty that takes us "out of the moment." It's not surprising that behind the beauty of music the ancients sensed mathematical mysteries waiting to be decoded. Among the great discoveries of the school of Pythagoras was the association of musical harmonies with simple ratios of numbers. For the ancients, this was a second clue that mathematics captures the patterns in the divine. We know few personal details about Pythagoras and his followers, but we can imagine that they noticed that an affinity for mathematics often accompanies a talent for music. We would say that mathematicians and musicians share an ability to recognize, create, and manipulate abstract patterns. The ancients might have talked instead of a shared ability to perceive the divine.

Galileo was exposed to music as a child, before he was a scientist.[3] His father, Vincenzo Galilei, was a composer and an influential music theorist, who is said to have stretched violin strings across the attic of their house in Pisa so his young son could experience the relationship between harmony and ratio. Bored during a service in the Pisa cathedral, Galileo noticed that the time it took a hanging lamp to sway from side to side was independent of how wide its swing was. This independence of the period (meaning the time it takes to complete one swing or orbit) from the amplitude of a pendulum was one of his first discoveries. How did he manage it? We would use a stopwatch or a clock, but Galileo didn't have those available. We can imagine that he simply sang to himself as he watched the lamps sway over his head, since he later claimed to be able to measure time to within a tenth of a pulsebeat.

Galileo evinced a musician's showmanship as well, when he took the case for Copernicanism to the people. He wrote his ideas down in Italian instead of Latin, the language of scholars, vividly conveying them through dialogues in which imagined characters converse about science as they share a meal or a walk. For this he is praised as a democrat who disdained the hierarchy of church and university to appeal directly to the intelligence of the common person.

But as brilliant a polemicist and experimenter as he was, what's stunning about Galileo's work are the new questions he asked — thanks in part to the liberation from ancient dogma that was the legacy of the Italian Renaissance. The ancient distinction between the earthly and divine realms that had long kept people from thinking seems to have left Galileo unimpressed. Leonardo had discovered proportion and harmony in static form, but Galileo looked for mathematical harmony in everyday motions, such as those of pendulums and balls rolling down inclined planes. Before he was democratic in his strategy of communication, he was a democrat about the universe.

Galileo destroyed the divinity of the sky when he discovered that heavenly perfection was a lie. He did not invent the telescope, and he may not have been the only one who used the new invention to look at the heavens. But his unique perspective and talents led him to make a fuss about what he saw there, which was imperfection. The sun has spots. The moon is not a perfect sphere of quintessence; it has mountains, just like Earth. Saturn has a strange threefold shape. Jupiter has moons, and there are vastly more stars than those seen with the naked eye.

This decline of divinity had been anticipated a few years earlier, in 1577, when the Danish astronomer Tycho Brahe watched a comet penetrate the perfect spheres of Heaven. Tycho was the last and greatest of the naked-eye astronomers, and he and his assistants accumulated over his lifetime the best measurements of planetary motions that had ever been made. These sat in his record books undecoded until 1600, when he employed an irascible young assistant, Johannes Kepler.

The planets move along the ecliptic, but they are not seen to move consistently. They all move in the same direction, but occasionally pause and reverse themselves, moving backward for a while. This retrograde motion was a great mystery to the ancients. Its real meaning is that the Earth is a planet, too, which moves around the sun as the other planets do. The planets appear to stop and start only from Earth's perspective. Mars moves eastward in our sky when it's ahead of us and reverses direction when Earth catches up. Its retrograde motion is simply an effect of Earth's motion, but the ancients couldn't see it that way, because they were stuck with the false idea that the Earth is at rest at the center of the universe. Since Earth is still, the perceived motion of the planets must be their real motion; hence the ancient astronomers had to explain the retrograde motions as if they were caused by the planets' intrinsic motion. To do so, they imagined an awkward arrangement involving two kinds of circles, in which each planet was attached to a small circle rotating around a point that itself moved on a bigger circle around the Earth.

The epicycles, as these mini-circles were called, rotated with a period of one Earth year, because they were nothing but the shadow of Earth's motion.[4] Other adjustments required still more circles; it took fifty-five circles to get it all to work. By assigning the right periods to each of the big circles, the Alexandrian astronomer Ptolemy calibrated the model to a remarkable degree of accuracy. A few centuries later, Islamic astronomers fine-tuned the Ptolemaic model, and in Tycho's time it predicted the positions of the planets, the sun, and the moon to an accuracy of 1 part in 1,000 — good enough to agree with most of Tycho's observations. Ptolemy's model was beautiful mathematically, and its success convinced astronomers and theologians for more than a millennium that its premises were correct. And how could they be

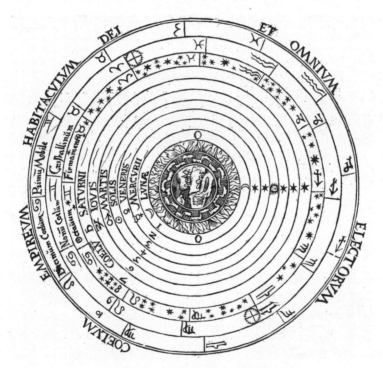

Figure 2: A schematic of the Ptolemaic vision of the universe.[5]

wrong? After all, the model had been confirmed by observation.

There's a lesson here, which is that neither mathematical beauty nor agreement with experiment can guarantee that the ideas a theory is based on bear the slightest relation to reality. Sometimes a decoding of the patterns in nature takes us in the wrong direction. Sometimes we fool ourselves badly, as individuals and as a society. Ptolemy and Aristotle were no less scientific than today's scientists. They were just unlucky, in that several false hypotheses conspired to work well together. There is no antidote for our ability to fool ourselves except to keep the process of science moving so that errors are eventually forced into the light.

It fell to Copernicus to decipher the meaning of the fact that all the epicycles have the same period and move in phase with the sun's orbit. He put Earth in its rightful place, as a planet, and the sun near the center of the universe. This simplified the model but introduced a tension the ancient cosmology couldn't survive. Why should Earth's sphere be any different from those of the heavens, if Earth is just another planet traveling through the heavens?

However, Copernicus was a reluctant revolutionary who missed other clues. A big one was that even after Earth's motion was accounted for, the planets' orbits were not precisely circles. Unable to escape the idea that motions on the sky must be compounded from circles, he solved this problem just as Ptolemy had fourteen centuries earlier. He introduced epicyles as needed to get the theory to fit the data.

The least circular orbit is that of Mars. It was Kepler's great luck — and science's too — that Tycho assigned to him the problem of deciphering the orbit of Mars, and, after working for many years after he left Tycho's service, Kepler found that Mars traces an ellipse, not a circle, in space.[6]

This was revolutionary in ways that may not be apparent to a modern reader. In an Earth-centered cosmology, the planets don't trace a closed path of any sort, because their paths relative to Earth each combine two circular motions with different periods. It is only when the orbits are plotted with respect to the sun that they make closed paths. Only then does it become possible to ask what the shape of an orbit is. So putting the sun at the center deepens the harmony of the world.

Once the planetary orbits were understood to be ellipses, the explanatory power of Ptolemy's theory was shattered. A slew of new questions arose: *Why* do the planets move in elliptical orbits? And what keeps them from wandering off? What compels them to move at all, rather than just sitting still in space? Kepler's answer was a wild guess that turned out to be half right: *What moves the planets around in their orbits is a force from the sun.* Imagine the sun as a rotating octopus, its arms sweeping the planets around as it turns. This was the first time anyone had suggested the sun as the source of a force that affects the planets. He just got the direction of the force wrong.

Tycho and Kepler smashed the heavenly spheres and in so doing unified the world. This unification had grave implications for the understanding of time. In the cosmology of Aristotle and Ptolemy, a timeless realm of eternal perfection surrounds the earthly realm. Growth, decay, change, all the evidence of a time-bound world is restricted to the small domain below the sphere of the moon. Above it is perfect circular motion, unchanging and eternal. Now that the sphere separating the time-bound and the timeless was smashed, there could be only one notion of time. Would this new world be time-bound throughout, with the whole universe subject to growth and decay? Or would timeless perfection be extended to all of creation, so that change, birth, and death would be seen as mere illusions? We still struggle with this question.

Kepler and Galileo did not solve the mystery of the relationship between the divine, timeless realm of mathematics and the real world we live in. They deepened it. They breached the barrier between sky and Earth, putting Earth in the sky as one of the divine planets. They found mathematical curves in the motions of bodies on Earth and the planets around the sun. But they could not heal the fundamental rift between time-bound reality and timeless mathematics.

By the middle of the 17th century, scientists and philosophers confronted a stark choice. Either the world is in essence mathematical or it lives in time. Two clues to the nature of reality hung in the air, expectant and unresolved. Kepler had discovered that the planets move along ellipses. Galileo had discovered that falling objects move along parabolas. Each was expressed by a simple mathematical curve and each was a partial decoding of the secret of motion. Separately they were profound discoveries; together they were the seeds of the Scientific Revolution, which was about to flower.

This is not unlike the present juncture in theoretical physics. We have two great discoveries, quantum theory and general relativity, whose unification we seek. Having worked on this problem for most of my life, I'm impressed by the progress we've made. At the same time, I'm certain that some simple idea lies hidden in plain sight that will be the key to its resolution. Admitting that progress can be held up as we await the invention of nothing more substantial than an idea is humbling, but it's happened before. The Scientific Revolution launched by the simple discoveries of Galileo and Kepler was long delayed because of the idea that the universe was divided into an earthly and a heavenly realm. This idea prevented the thorough application of mathematics to the lower world, while our understanding of the upper world was thwarted by the belief that there was no need to look for causes of perfect heavenly motions.

It's thrilling to think about what might have happened had this basic conceptual mistake not blinded, for more than 1,000 years, the thinking of smart people who had in their hands the data and mathematics needed to take the steps that Galileo did. A Hellenistic or Islamic astronomer could well have made some or all of Kepler's discoveries from data available 1,000 years earlier than Tycho. The idea that Earth orbits the sun did not have to wait for Copernicus; it was on the table ever since it was proposed by Aristarchus in the 3rd century BC. His heliocentric cosmology was discussed by Ptolemy and others and would have been known to such great scholars as Hypatia, a brilliant mathematician and philosopher who lived in Alexandria from about AD 360 to 415. Suppose she or one of her brighter students had discovered Galileo's law of falling bodies, or Kepler's elliptical orbits?[7] There might have been a Newton by the 6th century, and the Scientific Revolution might have started a full 1,000 years earlier.

Historians may protest that Copernicus, Galileo, and Kepler could not have made their discoveries before the Renaissance prepared the way by freeing thinkers from the dogmatism of the Dark Ages. But in Hypatia's time, the Dark Ages had not yet descended and the struggle between the exponents of Greek learning and religious fundamentalism had not yet killed the spirit of rational inquiry. History may have been quite different if someone in Roman Alexandria, or, for that matter, the great centers of learning that flourished in the Islamic world a few centuries later, had done away with the geocentric universe. However, the brightest scientists in the best conditions could not make the conceptual leap of imagining mathematical laws governing motion in the earthly sphere or dynamic forces playing a role in the heavens. It took the shattering of the spheres separating the two realms for Galileo and Kepler to make their discoveries.

But even they could not take the next step, which was to see the unity lying in the earthly parabola and the planetary ellipse. That took Isaac Newton.

Because they lived after the shattering of the spheres, Galileo and Kepler could have asked whether throwing something hard enough leads to orbiting and the slowing of an orbiting object leads to falling. To us, it's obvious that these are not two phenomena but one. But this was not apparent to them. Sometimes it takes a generation or so before the simplest implications of new discoveries come into focus. Half a century later, Newton understood that orbiting is a form of falling and completed the unification of the heavens and the Earth.

One clue was a mathematical unity shared by the two curves that code motion. Ellipses trace the planetary orbits and parabolas trace the paths of falling bodies on Earth. These two curves are closely related: They both can be made by intersecting a cone with a plane. Curves that can be so constructed are called conic sections; the other examples are circles and hyperbolas.

The question for the second half of the 17th century was to discover the physical unity explaining this mathematical unity. The insight that impelled Newton to embark on the Scientific Revolution was about nature, not mathematics, and it was not his alone. Several of his contemporaries had perceived the great secret: *The force that*

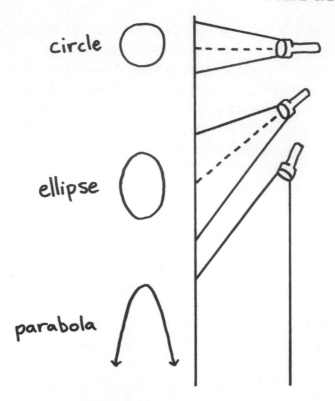

Figure 3: Conic sections illustrated with the image made by a flashlight on a wall.

impels everything on Earth to fall toward it is universal and acts also to pull the planets toward the sun and the moon toward the Earth. Gravity.

Newton, according to legend, had this epiphany while sitting in his garden noticing apples falling from a tree as he contemplated the motion of the moon. To complete the thought, he asked another crucial question: How does that force decrease with the distance between the objects? For decrease it must, otherwise we would be pulled upward to the sun rather than downward to Earth. And how does a force produce motion?

Others, such as Newton's contemporary, Robert Hooke, asked these questions, but the true accomplishment of Newton was in his answers to them. The effort took him two decades and resulted in the theory of motion and forces we call Newtonian physics.

For our purposes, the most important thing about these questions is that they're mathematical. How a force decreases with distance can be specified by giving a simple equation. The right answer, as every first-year physics student knows, is that the force decreases proportional to the square of the distance. The astounding consequence for our conception of nature is that such a simple mathematical relation captures a universal phenomenon in nature. Nature did not have to be so amazingly simple — and, indeed, the ancients had never contemplated such a simple and universal application of mathematics to the causes of motion.

To ask how a force causes motion, you have to think of a moving object tracing a curve in space. The question then is how the curve differs depending on whether there is a force acting on it or no force. The answer is stated in the first two of Newton's laws. If there is no force, the curve along which a body moves is a straight line. If there is a force, the force acts to cause an acceleration of the body.

It's impossible to state these laws without mathematics. A straight line is an ideal mathematical concept; it lives not in our world but in the Platonic world of ideal curves. And what is acceleration? It is the rate of change of velocity, which is itself the rate of change of position. To describe this adequately, Newton needed to invent a whole new branch of mathematics: the calculus.

Once you have the necessary mathematics, it's straightforward to work out the consequences. One of the first questions Newton must have answered with his new tools[8] was what path a planet would take under the influence of a force from the sun that decreases proportionally to the square of the distance. The answer: It can be an ellipse, a parabola, or a hyperbola, depending on whether the planets travel on a closed orbit or make a one-time pass by the sun. Newton was also able to subsume Galileo's laws of falling in his law of gravitation.[9] Galileo and Kepler had thus seen different aspects of a single phenomenon, which is gravity.

There is little in the history of human thought more profound than

the discovery of this hidden commonality between falling and orbiting. But beneath the enormity of Newton's accomplishment is an unintended consequence, which is that his work made our conception of
nature far more mathematical than before. Aristotle and his contemporaries had described motion in terms of tendencies: Earthly objects
have a tendency to seek Earth's center, air has a tendency to flee the
center, and so on. Theirs was an essentially descriptive science. There
is no suggestion that the paths along which objects move have any
special properties, and hence they had no interest in applying mathematics to the description of motion on Earth. Mathematics, being
timeless, was divine and applicable only to those divine and timeless
phenomena we could see, which were only in the heavens.

When Galileo discovered that falling bodies are described by a simple mathematical curve, he captured an aspect of the divine, brought it
down from the sky, and showed that it could be discovered in the motion of everyday objects on Earth. Newton demonstrated that the tremendous variety of motions on Earth and in the sky, whether impelled
by gravity or by other forces, are manifestations of a hidden unity. The
diverse motions are all consequences of a single *law of motion.*

By the time Newton had finished joining motion in the sky and on
Earth, we lived in a single, unified world. And it was a world infused
with divinity, because timeless mathematics was at the heart of everything that moved, on Earth and in the sky. If timelessness and eternity
are aspects of the divine, then our world — that is, the whole history of
our world — can be as eternal and divine as a mathematical curve.

3

A Game of Catch

To ADDRESS THE ISSUES raised in the first two chapters, we need to know more about how we define motion. Nothing seems simpler: Motion is change in position over time. But what is position, and what is time?

There are two answers that physicists have given to the seemingly innocuous question of defining position. The first is the commonsense idea that the position of an object is defined relative to a landmark of some sort; the second is that there is something absolute about position in space, beyond its relation to something else. These are called the relational and the absolute notions of space.

The relational notion of position is familiar to all of us. I am now three feet from my chair. The airplane is approaching the airport from the west and is now two kilometers from the end of runway 1 at a height of 1,000 feet. These are all descriptions of relative position.

But relational position seems to leave something out. Where is the ultimate reference? You give your coordinates on Earth, but where is Earth? So many miles from the sun, in the direction of the constellation Aquarius. But where is the sun? So many thousands of light-years from the center of the Milky Way Galaxy. And so forth.

Proceeding in this way, you can give the position of everything in the universe relative to everything else. This is a lot of information, but is it enough? Is there not some absolute notion of position — of where something *really* is, behind all these relative positions?

This debate between relational and absolute notions of space runs through the whole history of physics. Roughly speaking, Newton's physics was a triumph of the absolute picture, which was overthrown by Einstein's relativity theory, which established the relational view. I believe the relational view is correct, and I hope to convince the reader of this. But I would also like to give the reader a vivid sense of why savants like Newton embraced the absolute view and what is given up when we reject it in favor of the relational view.

To appreciate how Newton thought about the problem, we have to ask not only about position but about motion. Let's leave time aside for a moment and apply what we have just discussed. If position is relative, then motion is change of relative position — i.e., change of position relative to some reference body.

All commonplace talk of motion is talk of relative motion. Galileo studied bodies that fall relative to the surface of the Earth. I throw a ball and see it move away from me. The Earth moves around the sun. These are all examples of relative motion.

A consequence of relative motion is that who or what is moving is always a matter of point of view. Earth and the sun move around each other, but which is really moving? Is the real story that the sun moves around an Earth fixed at the center of the universe? Or is it rather the sun that is fixed, and Earth that orbits? If motion is only relative, there can be no right answer to this question.

The fact that anything can be moving or fixed makes it hard to explain the causes of motion. How could something be the cause of Earth's motion around the sun if there is a different and equally valid point of view according to which Earth isn't moving at all? If motion is relative, an observer is free to adopt the point of view that all motion is defined relative to him. To resolve the impasse and be able to speak of causes of motion, Newton proposed that there must be an absolute meaning to position. This was, for him, position with respect to what he called "absolute space." Bodies are moving or not, in an absolute sense, relative to this absolute space. Newton argued that it was the Earth and not the sun that moved absolutely.

The postulation of absolute space stops the infinite regression and gives a meaning to the location of every single thing in the universe, with no need to refer to anything else. This may be a comforting notion, but there's one problem. Where is this absolute space, and how would you measure the position of a body with respect to it?

No one has ever seen or detected absolute space. No one has ever measured a position that was not a relative position. So to the extent that the equations of physics refer to position in absolute space, they cannot be connected to experiment.

Newton knew this and it didn't bother him. He was a deeply religious thinker, and absolute space had a theological meaning for him. God saw the world in terms of absolute space, and that was enough for Newton. He would put it even more strongly: Space was one of God's senses. Things exist in space because they exist in the mind of God.

This isn't as strange as it sounds if you're a master decoder, as Newton was. He devoted years of work to searching for hidden meaning in the Scriptures, and as an alchemist he sought the hidden code for virtue and perhaps immortality. As a physicist, he uncovered universal laws that governed all motion in the universe but had previously been hidden. It was in character for him to believe that the essence of space was hidden from our senses yet seen by God.

Besides, he had a physical argument for absolute space. Even if position in absolute space could not be humanly perceived, some kinds of motion with respect to absolute space could be.

Children can't fly, but they can spin. And spin they do. Nothing matches the delight on the face of a child who has just discovered that she can make herself dizzy. Anytime she wants, over and over. Again! Newton had no children, but I like to imagine him being struck silent by the delight of his young niece, Catherine, spinning around in his study. Newton takes the wobbly, laughing child on his knee and tells her that her dizziness is a direct perception of absolute space. And absolute space is God. "What you feel when you feel dizzy is the hand of God upon you," he offers. She giggles, squirms as he starts to explain that she's dizzy not because she's rotating with respect to the furniture, or the house, or the cat, but because she spins with respect to space itself. And if space can make her dizzy, it must be something real. "Why?" she says, jumping off his lap to chase the cat out of the room. Let's leave Newton there, pondering gravity and mortality, and return to the question of how motion is defined.

When we say that something moves, we mean it is changing its position over time. This is common sense, but to be precise about it we need to be sure we know what we mean by time. Here we face the same dilemma of the relational versus the absolute.

Human beings perceive time as change. The time an event takes place is measured relative to other events — for example, the reading of the dial of a clock. All clock and calendar readings are relative times, just as addresses are relative positions. But Newton believed there is hidden behind change an absolute time, which God perceives.

Here's a taste of the debate that has raged over the issue of absolute time ever since. Newton's rival Gottfried Leibniz believed in God, too, but his God was not free, as Newton's was, to do as He pleased. Leibniz worshipped a supremely rational God. But if God is perfectly rational, then everything in nature must have a reason. This is Leibniz's principle of sufficient reason. One way to state it is that every question of the form "Why is the universe like this rather than like that?" must have a rational answer. There are, of course, questions for which there cannot possibly be any rational answer. Leibniz's point was that to ask a question that cannot have a rationally justified answer is to commit an error of thinking.

Leibniz illustrated his principle thus: He asked, "Why did the universe start when it did and not ten minutes later?" He replied that there cannot be any rational reason to prefer the history of the universe to one in which everything happens ten minutes later. All the relative times will be the same in both universes; only the absolute times are different. But the laws of nature speak only about relative times. Consequently, Leibniz argued, if there is no reason to prefer the universe to start at one absolute time rather than another, there can be no meaning to absolute time.

I accept Leibniz's reasoning, so whenever I refer to a time, I will mean a relative time. Indeed, although we can argue about whether there might be some transcendent sense in which absolute time exists, one thing that's certain is that we humans, living in the real world, have access only to relative times. So for the purpose of describing motions, we will consider time to be measured by clocks. For our purposes, a clock is any device that reads out a sequence of increasing numbers.

Now that we've defined both time and position, we can go on to measure motion: Motion is change of position, measured relative to some reference object, during a period of time, measured relative to the reading of a clock.

This brings us to the next, crucial step in our argument. To do science it is not enough just to make definitions and argue about concepts. You have to measure motions. This means using tools like clocks and rulers to associate positions and times with numbers.

Unlike absolute position, which is invisible, relative distances and relative times can be measured in numbers, which in turn can be recorded on a piece of paper or in a digital memory. In this way, observations of motion are turned into tables of numbers that can be studied with methods from mathematics. One such method is to make a graph of the records, thus turning the tables of numbers into pictures that can spark our understanding and imagination.

This powerful tool was developed by René Descartes and is taught to every schoolchild. It is undoubtedly something Kepler would have done as he struggled with Tycho's data on the orbit of Mars. In Figure 4 we see a graph of the orbit of the moon with respect to Earth.

In school, we learned a second way to draw a motion, which is to add an axis for time and graph the position against time. This repre-

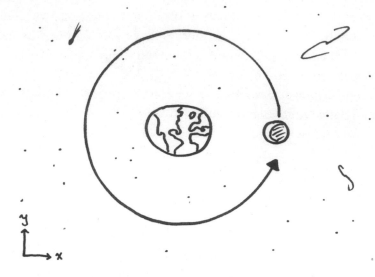

Figure 4: Graph of the moon's orbit around Earth.

sents the orbit as a curve in space and time, as in Figure 5. We see the moon's orbit now represented by a spiral; once it returns to its starting position, a month has passed.

Notice that by graphing records of observations, something wonderful has been done. The curve in Figure 5 represents measurements taken while something evolved in time, but the measurements themselves are timeless — that is, once taken, they do not change. And the curve that represents them is also timeless. By this means, we have made motion — that is, change in the world — into a subject of study by mathematics, which is the study of objects that don't change.

The ability to freeze time like this has been a great aid to science, because we don't have to watch motion unfold in real time; we can study the records of the past motions whenever we like. But beyond its usefulness, this invention has profound philosophical consequences, because it supports the argument that time is an illusion. The method of freezing time has worked so well that most physicists are unaware that a trick has been played on their understanding of nature. This trick was a big step in the expulsion of time from the description of

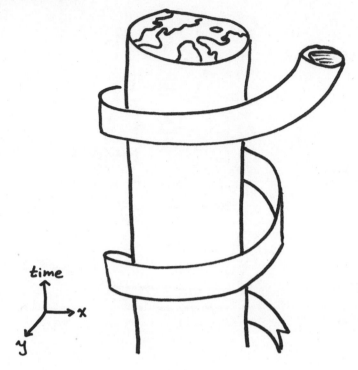

Figure 5: Graph of the moon's orbit as a curve in space and time.

nature, because it invites us to wonder about the correlation between the real and the mathematical, the time-bound and the timeless.

This correlation is so crucial that I want to frame it in an everyday example. All these weighty issues concern nothing more than what we can know about a game of catch.

❖

Around 1:15 P.M. on October 4, 2010, on the east side of High Park in Toronto, a novelist called Danny threw a tennis ball he had discovered that morning in his sock drawer to a poet he had just met, Janet.

To study Danny's throw through the eyes of physics, we do what Tycho and Kepler did for Mars. We observe the motion and record the ball's positions at a series of times; then we graph the results. To accomplish this, we need to give the position of the ball relative to some

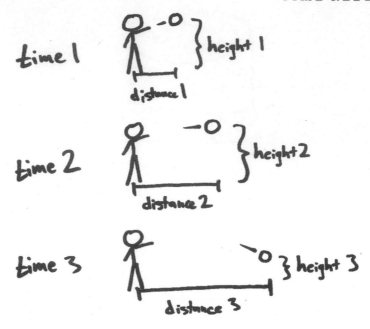

Figure 6: Danny's throw measured.

object, which we can take to be Danny himself. We also need a clock.

The ball moves fast, and this was a challenge for Galileo, but we can simply film Danny's throw and measure the position of the ball in each frame, along with the time of the frame. From the position of the ball in a frame we get two numbers, the ball's height off the ground and the horizontal distance the ball has traveled from Danny. (Space of course is three-dimensional, so we also have to describe the direction of Danny's throw. Other than to say he's throwing south, I'll ignore this complication here.) When we include the time of each frame, the record of the ball's trajectory is a series of triplets of numbers, one triplet for each frame of the film.

(time 1, height 1, distance 1)
(time 2, height 2, distance 2)
(time 3, height 3, distance 3)

And so on.

These lists of numbers are important tools if we are to study motion scientifically. But they are not the motion itself. They are just numbers, which are given meaning by the measurements we made of a ball in flight in a particular instance. The real phenomenon differs in several ways from the list of numbers that is its record. For example, many features of the ball have been neglected. We recorded only its positions, but the ball also has color, weight, shape, size, and composition. More important, the phenomenon unfolded in time: It happened just once and was gone, into the past. What's left is the record, and that is frozen, unchanging.

The next step is to graph the information in the record. Figure 7 is a picture of the path the ball made in space. We see that the ball flew on a parabola, just as Galileo predicted.

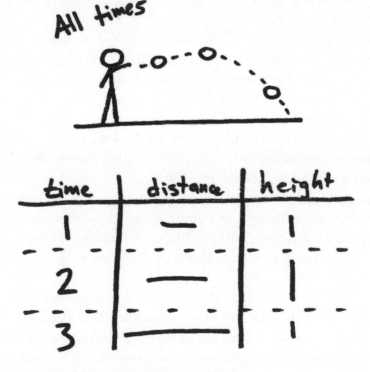

Figure 7: Danny's throw recorded and graphed.

We see again that the process of recording a *motion*, which takes place in time, results in a *record*, which is frozen in time — a record that can be represented by a curve in a graph, which is also frozen in time.

Some philosophers and physicists see this as a profound insight into the nature of reality. Some argue to the contrary — that mathematics is only a tool, whose usefulness does not require us to see the world as essentially mathematical. We can call these competing voices the *mystic* and the *pragmatist.*

The pragmatist will argue that there's nothing wrong with checking hypotheses about laws of motion by converting motion into numbers in tables and looking for patterns in those tables. But the pragmatist will insist that the mathematical representation of a motion as a curve does not imply that the motion is in any way identical to the representation. The very fact that the motion takes place in time whereas its mathematical representation is timeless means they aren't the same thing.

Some physicists since Newton have embraced the mystic's view that the mathematical curve is "more real" than the motion itself. The great attraction of the concept of a deeper, mathematical reality is that it is timeless, in contrast to a fleeting succession of experiences. By succumbing to the temptation to conflate the representation with the reality and identify the graph of the records of the motion with the motion itself, these scientists have taken a big step toward the expulsion of time from our conception of nature.

The confusion worsens when we represent time as an axis on a graph, as we did in Figure 5. In Figure 8, we see the information about the trajectory of Danny's ball including clock readings, displayed as if they were measurements made by a ruler. This can be called spatializing time.

And the mathematical conjunction of the representations of space and time, with each having its own axis, can be called *spacetime.* The pragmatist will insist that this spacetime is not the real world. It's entirely a human invention, just another representation of the record we

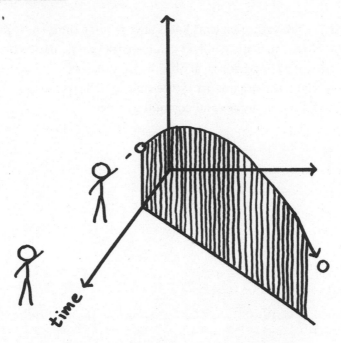

Figure 8: Danny's throw graphed as a curve in space and time.

have of the process of Danny throwing the ball to Janet. If we confuse spacetime with reality, we are committing a fallacy, which can be called the fallacy of the spatialization of time. It is a consequence of forgetting the distinction between recording motion in time and time itself.

Once you commit this fallacy, you're free to fantasize about the universe being timeless, and even being nothing but mathematics. But, the pragmatist says, timelessness and mathematics are properties of representations of records of motion — and only that. They are not, and cannot be, properties of real motions. Indeed, it's absurd to call motion "timeless," because motion is *nothing but* an expression of time.

There's a simple reason that no mathematical object will ever provide a complete representation of the history of the universe, which is that the universe has one property no mathematical representation of it can have. Here in the real world, it is always some time, some present moment. No mathematical object can have this particularity, because, once constructed, mathematical objects are timeless.[1]

Who is right, the pragmatist or the mystic? This is the question on which the future of physics and cosmology turns.

4

Doing Physics in a Box

IN HIGH SCHOOL, I was cast in Jean-Paul Sartre's *No Exit*. I played Joseph Garcin, a man locked in a small room with two women, all of us deceased. The play was an extreme version of a society in a box; it enabled the playwright to examine the consequences of our moral choices. In the climactic scene, I was to bang on the classroom door, shouting the famous line, "Hell is other people!" but the door's pane shattered, showering me with glass and bringing my acting career to a close.

Musical performance, like theater, allows a heightened examination of human emotion by isolating us in a controlled environment. As a teenager, I listened to a terrifying performance of my cousin's band, Suicide, in the basement of the Mercer Arts Center in Greenwich Village. The singer locked the doors and mesmerized the audience with a long aria of wanton murder, sung over a numbing repetition of the chords of the garage-rock classic "96 Tears." The atmosphere grew claustrophobic as the singer became increasingly menacing, but like the characters in *No Exit*, we were stuck. More recently, the insight-through-claustrophobia method has been adopted by conceptual artists who lock unlikely couples — like an artist and a scientist — in a room for twenty-four hours at a time and videotape everything that happens.[1]

In both play and performance, the isolation is fake. With enough effort, anyone could simply walk out at any time. We don't, because there's much to be learned from subjecting ourselves to the rigors of a reduced social environment. Less really is, in this sense, more. Art seeks universals through detailed examination of particulars[2], which often requires an artificially restricted setting to succeed.

It's the same with physics. Most of what we know about nature has come from experiments in which we artificially mark off and isolate a phenomenon from the continual whirl of the universe. We seek insight into universals of physics through restricting our attention to the simplest of phenomena. The method of restricting attention to a small part of the universe has enabled the success of physics from the time of Galileo. I call it *doing physics in a box.* It has great strengths but several weaknesses, and both are essential to our story of the expulsion of time from physics and its rebirth.

We live in a universe that is always changing, full of matter that is always moving. What Descartes, Galileo, Kepler, and Newton learned to do was to isolate little pieces of the world, examine them, and record the changes in them. They showed us how to display the records of these motions in simple diagrams whose axes represent the positions and times in a way that is frozen and hence amenable to being studied at our leisure.

Notice that to apply mathematics to a physical system, we first have to isolate it and, in our thinking, separate it out from the complexity of motions that is the real universe. We couldn't get very far with the study of motion if we worried about how everything in the universe affects everything else. The pioneers of physics, from Galileo to Einstein and up to the present day, could make progress because they could isolate a simple subsystem, like a game of catch, and study how the ball moves. In reality, though, a ball in flight is influenced in a myriad of ways by things outside the subsystem we defined. The simple description of a game of catch as an isolated system is a crude approximation of the real world — although one that has proved fruitful in discovering fundamental principles that appear to govern all motion in our universe.[3]

This kind of approximation, in which we restrict our attention to a few variables or a few objects or particles, is characteristic of doing physics in a box. The key step is the selection, from the entire universe, of a subsystem to study. The key point is that this is always an approximation to a richer reality.

It's easy to generalize our treatment of the game of catch to a large number of systems we study in physics. To study a system, we need to define what it contains and what is excluded from it. We treat the system as if it were isolated from the rest of the universe, and this isolation is itself a drastic approximation. We cannot remove a system from the universe, so in any experiment we can only decrease, but never eliminate, the outside influences on our system. In many cases, we can do this accurately enough to make the idealization of an isolated system a useful intellectual construct.

Part of the definition of a subsystem is a list of all the variables we need to measure in order to determine everything we want to know about it at a moment of time. The list of these variables makes up an abstraction we call the *configuration of the system.* To represent the set of all possible configurations, we define an abstract space called the configuration space. Each point of the configuration space represents one possible configuration of the system.

The process by which the configuration space is defined starts with extracting the subsystem from the larger universe. Hence, the configuration space is always an approximation to a deeper and more complete description. The configuration and its representation in a configuration space are both abstractions — human inventions that are helpful for the method of doing physics in a box.

To describe a game of pool, we can choose to record the locations of sixteen balls on a two-dimensional table. It takes two numbers to locate a single ball on the table (its position relative to the table's length and width), so the full configuration will require a list of thirty-two numbers. The configuration space has one dimension for each number that must be measured, so in the case of pool, it's a thirty-two-dimensional space.

But a real pool ball is an immensely complicated system, so its representation as a single object with a single position is a drastic approximation. If you want a more precise description of a game of pool, you should record the positions not just of the balls but of every atom in every ball. This requires at least 10^{24} numbers, hence a configuration space of that high dimension. But why stop there? If a description at the level of atoms is what's wanted, you should include the positions of all the atoms in the table, all the atoms in the air that impinge on the balls, all the photons that light up the room — and then why not all the atoms in the Earth, sun, and moon attracting the balls gravitationally? Anything less than a cosmological description will be an approximation.

Something else left outside the subsystem is a clock, which we use to specify the moment in which an observation is made. The clock is not considered part of the subsystem, because it's assumed to tick uniformly despite whatever is going on in the subsystem. The clock provides us with a standard against which the subsystem's motion is to be recorded.

The use of an external clock violates the concept that time is relational. Change in the system is measured by reference to the external clock, but nothing that happens in the system is meant to influence the external clock. This is convenient, but it's allowed only because we're making a rough approximation, in which we neglect all the interactions between the system and everything outside it, including the clock.

If we take this method too seriously, we may be tempted to imagine a clock external to the whole universe, by which we can measure change in the universe. This is the route to a big conceptual mistake, which is to believe that the universe as a whole evolves with respect to some absolute notion of time coming from outside it. Newton made this mistake because he was caught in the fantasy that the physics he invented captured God's view of the universe as a whole. The mistake persisted until Einstein corrected it — by finding a way, within relativ-

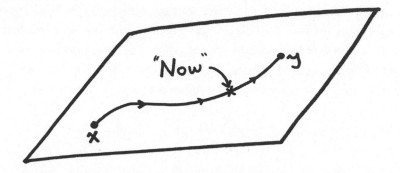

Figure 9: Configuration space and a history through it. The X marks a moment of time.

ity theory, to put the clock inside the universe — and we should not make it again.

However, as long as we don't take it too seriously, the picture of a small subsystem of the universe evolving as measured against the readings of an external clock is a useful approximation. At each time we measure, we get a list of numbers characterizing the configuration at that time, which then defines a point in the configuration space. By imagining the time measurements as rapid-fire, we can idealize this set of points to a curve through the configuration space (see Figure 9). This represents a history of the subsystem as captured by a sequence of measurements of its configurations.

Just as in the case of Danny's ball in his game of catch with Janet, time has disappeared from the picture. What's left is a trajectory through the space of possible configurations. This trajectory is a curve summarizing information in a set of records of something that happened in the past. When we're done, we have a representation of the motion of the subsystem — motion that unfolded in time, just once — by a timeless mathematical object, which is a curve in the space of possible configurations of the subsystem.

The configuration space is timeless; it is assumed to just be there, forever. When I refer to it as "the space of possible configurations," I mean that if I wished, I could put the subsystem into any one of those configurations at any time. The history of the system is then represented by a curve starting at that initial configuration. That curve, once drawn, is timeless. This brings us back to our key question: Is the disappearance of time in the representation a deep insight into the nature of reality, or is it a misleading and unintended consequence of a method for approximately describing small parts of the universe?

◆

Newton did more than invent a way to describe motion; he invented a way to predict it. Galileo discovered that in the case of a thrown ball, the curve is a parabola. Newton gave us a method for determining what the curve would be in an enormous variety of cases. This is the content of his three laws of motion. His laws can be summarized as follows:

To predict how a ball will travel, three pieces of information are needed:

- The ball's initial position;
- the ball's initial velocity (that is, how fast and in which direction it's moving);
- the forces that will influence the ball as it moves.

Given these three inputs, Newton's laws of motion can be used to predict the future path of the ball. We can program a computer to do this for us. Give it the three inputs, and it will output the path the ball will follow. This is what we mean when we speak of a "solution" to Newton's laws. A solution is a curve in configuration space. It represents a history of the system, from the moment the system is prepared or first observed onward. That first moment is called the initial condition. You describe the initial condition when you give the initial position and velocity. The laws then kick in and determine the rest of the history.

One law has an infinite number of solutions, each of which describes a possible history of the system in which the laws are satisfied. You specify which history describes a particular experiment when you give the initial conditions. Thus, to predict the future or explain something, it's not enough to know the laws; you must also know the initial conditions. In laboratory experiments this is easy, because the experimenter prepares the system to start in some particular initial condition.

Galileo's law of falling bodies says that the ball Danny threw will trace a parabola. But which parabola? The answer is determined by how fast and at what angle and from what location he threw the ball — that is, by the initial conditions.

It turns out that this method is general. It can be applied to any system that can be described by means of a configuration space. Once the system is specified, the same three inputs are needed:

- *The initial configuration of the system.* This gives a point in the configuration space.
- *The initial direction and speed of the changes of the system.*
- The forces the system will be subject to as it changes in time.

Newton's laws then predict the precise curve in configuration space that the system will follow.

The generality and power of Newton's method cannot be underestimated. It has been applied to stars, planets, moons, galaxies, clusters of stars, clusters of galaxies, dark matter, atoms, electrons, photons, gases, solids, liquids, bridges, skyscrapers, cars, airplanes, satellites, rockets. It has been applied usefully to systems with one, two, or three bodies and systems with 10^{23} or 10^{60} particles. It has been applied to fields — such as the electromagnetic field — whose definition requires the measurement of an infinite number of variables (the electric and magnetic fields at each point in space, for example). It has described an enormous number of possible forces or interactions among the variables that define the system.

The basic method can also be applied in computer science, where it's called the study of cellular automata. And with only a little modification, it's the basis for quantum mechanics.

Because of the power of this method, it can be called a paradigm.

We will name it after its inventor: the *Newtonian paradigm.* This is a more formal way to talk about the method of doing physics in a box.

In its essentials, the Newtonian paradigm is constructed from the answers to two basic questions:

- What are the possible configurations of the system?
- What are the forces that the system is subject to in each configuration?

The possible configurations are also called the *initial conditions,* because we specify them to get started. The rules by which the forces and their effects are described are called the *laws of motion.* These laws are represented by equations. When you put the initial conditions into the equations, the equations give you the future evolution of the system. This is called solving the equations. There are an infinite number of such solutions, because there are an infinite number of possible initial conditions.

We should be aware that this powerful method is based on some powerful assumptions. The first is that the configuration space is timeless. It's assumed that some method can give the whole set of possible configurations ahead of time — that is, before we watch the actual evolution of the system. The possible configurations do not evolve, they simply are. A second assumption is that the forces, and hence the laws the system is subject to, are timeless. They don't change in time, and they also presumably can be specified ahead of the actual study of the system.

The lesson here is as simple as it is terrifying. To the extent that the assumptions underlying the Newtonian paradigm are realized in nature, time is inessential and can be removed from the description of the world. If the space of possible configurations can be specified timelessly, and the laws can as well, then the history of any system need not be seen as evolving in time. It is sufficient, for answering any question physics can pose, to see the whole history of any system as a single frozen curve in configuration space. The seemingly most essential aspect of our experience of the world — its presentation to us as a succession of present moments — is missing from our most successful paradigm for the description of nature.

We began by watching a tennis ball, with a phone number on the side in purple ink, thrown between two writers named Danny and Janet, in the afternoon of October 4, 2010, in High Park. Our deepest understanding of how it moved amounts to contemplating a timeless picture containing a colorless curve in an abstract space.

5

The Expulsion of Novelty and Surprise

THE INVENTION OF THE Newtonian paradigm as a general method for doing physics in a box was a key step in the expulsion of time. One consequence was the argument for determinism famously articulated by Pierre-Simon Laplace, who claimed that if he were given the precise positions and motions of all the atoms in the universe, together with a precise description of the forces they were subject to, he could predict the future of the universe with total accuracy. This statement has convinced many since that the future is completely determined by the present.

But there is one important assumption of the argument that can be questioned, which is that you can extend the Newtonian method to the universe as a whole, by including everything in the box. But physics in a box starts by isolating a small subsystem of the universe. Can Laplace really get away with ignoring that step?

Let's go back to the game of catch in the park.

It's now August 14, 2062, at 3:15 in the afternoon. Laura, the granddaughter of Danny and Janet, will be throwing a Frisbee to Francesca, the daughter of Billy and Roxanne, who also grew up by the park. As

Laura flings the Frisbee, Francesca is distracted by the flash of a message coming into the microcellphone implanted in her retina. Does she miss the Frisbee?

If you believe that the Newtonian paradigm applies exactly to the world, then you believe that it was already determined in 2010 whom Danny and Janet will marry (each other, as it happens, but neither would have guessed that at the time), when their son will be conceived, who he will marry, and when his daughter will be conceived, and whether or not she will have a predilection for Frisbee. You have to believe that every single motion, thought, idea, and emotion that these people will ever have is already determined in the present. You have to believe that the entire list of all those who will ever live is already set down, even if it's impossible to imagine the technology to decode it.

You have to believe not just that it is already determined — and, indeed, has been for billions of years — that there will be a Frisbee game between Laura and Francesca that afternoon, even though they grew up on opposite sides of the park and met not five minutes before. And you have to believe that nothing could be done now to prevent the development of retinal-implantable microcellphones, and that nothing could be done to prevent that fateful message from being sent at exactly that moment and distracting Francesca. Does she nonetheless catch the Frisbee? Before her microphone flashes, no one watching could know, but if the future is determined there is, in principle, already some quantity that could be measured now that would tell us.

The claim that the laws of physics, plus the initial conditions, determine the future down to the smallest detail is an astounding claim, because in the long run the smallest details do matter. In each successful conception, one out of roughly 100 million sperm impregnates the egg. This has happened roughly 100 billion times since there were humans and trillions of times before that, during the evolution of our ancestors. Trillions of choices of one out of 100 million is an awful lot of information, but we have to believe that all this, and much, much more, was written into the initial conditions of the universe at some much earlier time. And this is just one small detail of life on one small planet.

This is part of the meaning of the statement that in the Newtonian paradigm time disappears. All the things that have ever happened, that are happening now, and that will ever happen are just points on a trajectory in the configuration space of the universe, a curve that is already determined. The passage of time brings no novelty or surprise, for change is just a rearrangement of the same facts.

If there is to be a space for novelty and surprise, there must be something wrong with the Newtonian paradigm, or at least with extending it from a method to study small subsystems of the universe to an exact description of the whole universe. One limitation is that if the future is determined given the initial conditions, you need to know what determines the initial conditions. As you seek the reasons for things to be as they are and not otherwise, you are driven deeper and deeper into the past.

As you go further into the past, you have to consider a larger and larger region of space, containing events that might have influenced any ancestor of Danny or Janet. If you go back a million years to the chance meeting of two of their *Homo erectus* ancestors from different nomadic groups, you have to survey a region 2 million light-years across to assure yourself that there's no supernova close enough to do damage to the Earth. If we go all the way back to the origin of life on Earth, we have to survey a good fraction of the observable universe.

Thus, if we seek not just necessary but sufficient causes, we cannot avoid the conclusion that the full set of sufficient causes of Danny's meeting Janet include conditions at cosmological distances and times from that happy event. As we keep pushing the chain of causes back, sooner or later the whole universe is involved. And before we get to the end of the causes, we find ourselves at the moment of the Big Bang. So the ultimate sufficient causes of Danny and Janet's meeting are in the initial conditions of the universe at the Big Bang. The ultimate applicability of the argument for determinism is thus a question about cosmology. If we want to understand whether and how their meeting was determined, we need a theory of the universe as a whole.

The problem of determinism collides with the fact that the method of doing physics in a box applies to small subsystems of the universe.

Before we can answer the question of whether seemingly chance events in our lives are determined completely by past conditions, we have to know whether our theories can be scaled up to theories of the complete universe.

We live in a world in which the flap of a butterfly's wing can influence the weather oceans away and months later. In general terms, small changes in initial conditions are amplified exponentially to big changes in outcomes. This is why doing physics in a box necessarily involves approximations. These include the selection we make of the observable quantities to model in the configuration space and the neglect of the influence of everything else in the world on them.

You can easily imagine filling in these details, however. If you know the laws of physics applicable to the smallest particles that make up the subsystem, then you can at least imagine making an exact description of all the variables needed to describe the subsystem and all the forces by which these variables interact. The most precise description we have currently of the laws of nature and the elementary particles is the Standard Model of Particle Physics, which is easily situated within the Newtonian paradigm. This model contains all we know of nature, except for gravity, and it has repeatedly survived various experimental tests.

So, why not put in the rest of the universe? You can imagine treating a larger subsystem that contains our system — not just Danny's tennis ball but everything and everyone in the park that afternoon. Expand that again to include everyone and everything in the city of Toronto; expand it again to include everything on and inside the Earth and within a million miles of it. Each time you expand the subsystem, you can still use the same laws of physics — hence you can employ the Newtonian paradigm. In each case, the approximation gets better and better and so does the strength of the argument for determinism.

But there's always something left out. Just beyond the solar system there could be a big black cloud that will swallow the sun in a year, or a comet that will graze the Earth in ten years. These events could disrupt the coming marriage of Danny and Janet. The perturbation doesn't have to be large or involve Earth directly. Danny's attention could have been captured by a news story about a comet coming close to Jupiter, and he went to the park a minute later and never met Janet. Millions of people who would have been their descendants will never live. In our world, the amplification of small events into big consequences is the normal state of affairs.

A deterministic physical theory can be likened to a computer. The configuration space is the memory, into which data is put. The law is analogous to the program. You run the program and it acts on the input data, which is then overwritten with the output data. Given the input and the program, the output is completely determined. Every time you run it with the same input, you get the same output. But here's something else to think about: The output is determined from the input and program in two rather different ways.

If we consider the computer as a physical device, then it operates according to the laws of physics. From this point of view, the output is causally determined by the input. It's the result of laws of physics acting on initial conditions. This process requires time, because the causal process, as directed by the laws of physics, is carried out in time.

But the output is determined in another way, too. The output is *logically* implied by the input and the program. The input and output are representatives of mathematical objects. The program is also a mathematical object. You could prove logically that the output is a mathematical consequence of the combination of the input and the program. This logical determination requires no time, because no physical process is involved. The proof that the output is implied by the input and the program is a mathematical fact, which lives in the timeless world of true facts about mathematical objects.

This is the sense in which time is removed from the description of physics within the Newtonian paradigm. It is unnecessary to actually run the computer to know what the output is, because the output can be deduced by a sequence of logical arguments. How the deduction is carried out is irrelevant; the computer is just a tool exploiting the laws of physics to model a logical implication through a causal process. But there are an infinite number of possible ways the computer might be built and programmed that would lead to exactly the same outcome.

The point is that there is no information in the output that is not already logically implied by the input. The output is just a rearrangement of the input according to some logical rule. This is the sense in which nothing novel or surprising can ever be produced. Nor is there any need for causal evolution to act out in time just to reproduce the effect of a logical, and hence timeless, implication.

The same holds for any system described within the framework of the Newtonian paradigm. In all such cases, the final configuration is just the result of the laws of physics acting on the initial conditions. The configuration space where the initial configuration and the final configuration live is a mathematical object, as are those configurations. Once the laws are expressed as mathematical equations, the evolution of the initial conditions into the final configuration after a certain amount of time is a mathematical fact. It can be deduced mathematically; in fact, it can be proved as a mathematical theorem. What the Newtonian paradigm does is replace causal processes — processes playing out over time — with logical implication, which is timeless. This is yet another way that time is eliminated by the Newtonian paradigm.

❖

One way to see that surprise and novelty can play no role is to consider that a law of physics can often be run in reverse. If you think of a law of physics as a computer or machine that turns initial conditions into a final configuration, then you can imagine the law as having a toggle switch that can be flipped to reverse the direction of time. What you do is flip the switch and insert the final configuration into the input. You run the law for the same amount of time as before, except that this time the law will run backward to turn the final configuration back into the initial one. We say that a law that can be run backward to turn every final configuration into its initial conditions is time-reversible.

Let's look at a simple example: the motion of Earth as it spins on its axis and orbits the sun. Reversing the direction of time reverses the direction of the orbit and the spin of the Earth, but such an orbit is also permitted by Newton's laws. If you took a movie of the Earth's motion and showed it to aliens, they would say (if they had any conception of laws) that Newton's laws are governing the motion. But the same would be true if you gave them a print of the film run backward; they would conclude that this was an orbit allowed by Newton's laws. Indeed, if you gave them both films and asked them to say which was the original and which the reversal, they couldn't tell. The same would be true of filmed motions of the whole solar system, with the eight planets and myriads of other bodies.

Of course, many of us have seen films run backward and most look strange or funny. Often it's not because the reversed motion would be impossible according to the laws of physics; rather, the motion is possible but exceedingly improbable. This is generally true of complex systems involving large numbers of such things as atoms. Here we must deal with the laws of thermodynamics, which are not reversible in time and which I will discuss in chapters 16 and 17.[1] For the present, let's consider just two simple examples.

Many laws of physics are time-reversible. One is Newtonian mechanics, another is general relativity, still another is quantum mechanics. The Standard Model of Particle Physics is almost time-reversible but not fully so. (There is one mostly inconsequential aspect of the weak nuclear interaction that does not reverse.) If you take a history that evolved according to the Standard Model, reverse the direction of time and simultaneously make two other changes, you get another history that the model allows. Those two changes are replacement of the particles by their antiparticles and reversing left and right. The operation is called CPT (for charge, parity, and time reversal), and you can think of it as a different way of running a film backward. Any theory consistent with quantum mechanics and special relativity allows the direction of time to be reversed in this way.

These reversals are another argument for the unreality of time. If the direction of the laws of nature can be reversed, then there cannot, in principle, be any difference between the past and the future — and the fact that we have very different relationships with the past and the future cannot be a fundamental property of the world. The apparent differences between the future and the past must either be illusions or consequences of special initial conditions.

Ludwig Boltzmann, who, with his insights into the nature of entropy, did more than anyone else to connect the atomic world with the macro world we experience, once said, "For the universe, the two directions of time are indistinguishable, just as in space there is no up and down."[2] And if there's no real distinction between past and future — if they have exactly the same content, just logically rearranged — then there's no need to believe in the reality of the present moment or the reality of the passage of time. The time-reversibility of the laws of physics is often taken as another step in the removal of time from the physicists' conception of nature.

We have just a few more steps to take before time is gone completely from physics. The next step comes from Einsteinian relativity, which will provide us with the strongest argument of all for the unreality of time.

6

Relativity and Timelessness

W HEN I WAS NINE YEARS OLD, my father brought a copy of Lincoln Barnett's *The Universe and Dr. Einstein* home to our apartment on Manhattan's Upper West Side, and we puzzled together over its explanations of relativity theory. Even now I can recall the diagrams of speeding trains and bent starlight. This was my first exposure to physics.

Then, around age sixteen, I read Einstein's first paper on general relativity while riding the subway downtown to visit my rock-band-member cousin. Einstein's seminal papers were available then, as they are now, in a cheap paperback edition.[1] Seduced into physics by his writing, which I was fortunately exposed to before I had the chance to open a textbook, I had no idea that I'd encountered the best example of how to express in clear ideas the essence of nature. It was something like being weaned off infant formula in a five-star French restaurant, so that later you would have to be coerced, kicking and screaming, into eating your Cheerios and peanut-butter-and-jelly.

Later I discovered that there's very little in physics to match the conceptual clarity and elegance of Einstein's theories. Not quantum mechanics, not contemporary quantum field theory, not even New-tonian mechanics, whose textbook presentations are often logical messes, undermined by confused and circular definitions of basic con-cepts such as mass and force. But because I began with Einstein, his work became my scientific standard and his theories of relativity be-came my touchstones, their principles as sacred as any text could be to one schooled in the skepticism of science.

As it happens, Einstein's theories of relativity are the strongest argu-ments we have for time being an illusion masking a truer, timeless uni-verse. Back when I believed that time is an illusion, my main reasons had to do with relativity theory.

Einstein invented two theories of relativity. The first, special relativ-ity, is a theory about a world in which gravity does not exist. It was pre-sented in two of the papers Einstein published in his "miracle year" of 1905.[2] General relativity, invented over the next decade, incorporates gravity.

Einstein's two theories of relativity are, at their most basic, theories of time — or, better, timelessness. They have an undeserved reputa-tion for being difficult; I find them beautifully simple and easy to ex-plain. It's true that relativity seems at first counterintuitive, because it replaces a wrong intuition by a deeper intuition that experiment tells us is closer to the truth. To learn relativity is to experience a transition from one way of mentally organizing the world to another. You have to give up certain unconscious assumptions about time, but after that the main ideas follow logically.

In this chapter, I will talk only about the ideas and results of relativ-ity theory that bear on the nature of time. I will make assertions that I hope are clear, but I won't do what's usually done in popular physics books, which is to give the arguments that connect Einstein's simple postulates to their counterintuitive consequences.[3]

We'll concern ourselves with two concepts from special relativity. The first is the *relativity of simultaneity.* The second, which follows from it, is the *block universe.* Each was a major step in the expulsion of time from physics.

In crafting special relativity, Einstein brought two strategies to bear on the question of the nature of time. First, he embraced the relational side of the debate about whether time is relational or absolute: Time is about change, which means it's about perceived relationships. There's no such thing as an absolute or universal time.

In his early work, Einstein also utilized a strategy called *operationalism.* According to this approach, the only meaningful way to define a quantity like time is to stipulate how to measure it. If you want to talk about time, you must describe what a clock is in your theory, and how the clock works. When you're approaching science operationally, you ask not about what is real but about what an observer could observe. The observer's situation in the universe must be taken into account, including where she is and how she's moving. This enables you to ask whether different observers will agree or disagree about what they're seeing. Some of Einstein's most interesting discoveries are about what observers disagree about.

But what about reality? Aren't physicists interested in what is real rather than just what is observed? Yes, but while most operationalists believe in reality, they also believe that the only way to get at it is through what is observed. The test of whether something is real — objectively true — is that all observers will agree on it.

The great discovery about time in Einstein's theory of special relativity is called the relativity of simultaneity. It has to do with whether two events distant from each other can be considered as taking place at the same time. What Einstein found was that there's an ambiguity in any definition of distant events as simultaneous. Observers in motion with respect to each other will reach different conclusions about whether two events are simultaneous or not when those two events are distant from each other.

It's perfectly natural for someone waking up in Toronto to wonder what her lover is doing at the same moment in Singapore. If this makes sense, then it also must make sense to ask what's happening at that moment on Pluto, on a planet in the Andromeda Galaxy, or, indeed, throughout the universe. What Einstein showed is that our natural intuition that it's meaningful to talk about what's happening right now far from us is mistaken. Two observers who move with respect to each other will disagree about whether two distant events are simultaneous.

The relativity of simultaneity depends on some assumptions, one of which is that the speed of light is universal — which means that any two observers who measure a photon's speed will agree on their measurements, no matter how they're moving with respect to each other or the photon. We also can assume that nothing can travel faster than this universal speed.[4] Given this, an event can influence another event only if a signal traveling at the speed of light or less leaves the first and arrives at the second. If this could happen, we say that the two events are *causally related,* in that the first could be a cause of the second.

But two events can be so far apart in space and take place so close in time that no signal can reach from one to the other. In such cases, neither of the two events can be the cause of the other. We say that two such events are not causally related. Einstein showed that in these cases you can't state whether they're simultaneous or one happened before or after the other. Both answers are possible, depending on the motion of the observers carrying the clocks by which time is measured.

For physics to make sense, observers have to agree on the order of causally related events to avoid confusion about the attribution of causes. But there's no reason for observers to agree about the order of events that could not possibly affect each other. In Einstein's theory of special relativity, they don't agree.

So it makes no sense for our friend in Toronto to wonder what her lover is doing *right now* in Singapore.[5] But it makes total sense for her to think about what he was doing a few seconds ago. Those seconds are more than enough time for him to have sent the text she is reading now; his sending the text and her reading it are causally related events. And all observers will agree that the text she sends now will change the rest of his life, beginning with when he reads her news a minute later.

Besides the existence of a universal speed limit that all observers agree on, special relativity depends on one other hypothesis. This is the *principle of relativity* itself. It holds that speed, other than the speed of light, is a purely relative quantity — there's no way to tell which observer is moving and which is at rest. Suppose two observers approach each other, each moving at a constant speed. According to the principle of relativity, each can plausibly declare herself at rest and attribute the approach entirely to the motion of the other.

So there's no right answer to questions that observers disagree about, such as whether two events distant from each other happen simultaneously. Thus, there can be nothing objectively real about simultaneity, nothing real about "now." The relativity of simultaneity was a big blow to the notion that time is real.

What observers do agree about can be called the *causal structure.* Pick any two events in the history of the universe and call them X and Y. Then one of three things will be true. Either X could be a cause of Y, or Y could be a cause of X, or neither could be a cause of the other. These causal relations are agreed to by all observers. The causal structure is the list of all these relations, for all events in the universe. Thus you can say that what is physically real in the history of the universe includes its causal structure.

This is a timeless picture, because it refers to the whole history of the universe at once. There is no preferred moment of time, no reference to what time it is now, no reference at all to anything corresponding to our experience of the present moment. No meaning to "future" or "past" or "present."

If you remove everything corresponding to the observations of particular observers from the description of nature given by special relativity, there remains the causal structure. Since this is all that's observer-independent, it must—if the theory is true—correspond to physical reality. Hence, to the extent that special relativity is based on true principles, the universe is timeless. It is timeless in two senses: There is nothing corresponding to the experience of the present moment, and the deepest description is of the whole history of causal relations at once.

This picture of the history of the universe given by causal relations realizes Leibniz's dream of a universe in which time is defined completely by relations between events. Relationships are the only reality that corresponds to time—relationships of a causal sort.

Actually, besides the causal structure, there's another piece of information that observers agree about. Consider a physical clock, which ticks off seconds, floating freely in space. It strikes noon, then a minute later it strikes a minute past noon. The first event can be considered a cause of the second. In between, the clock ticked sixty times. The number of times it ticked between the two events is something else all observers, regardless of their relative motion, can agree about. This is called the *proper time.*[6]

The picture of the history of the universe, taken as one, as a system of events connected by causal relations, is called the *block universe.* The reason for that perhaps peculiar name is that it suggests that what is real is the whole history at once — the allusion is to a block of stone, from which something solid and unchanging can be carved.

The block universe is the culmination of the movement begun by Galileo and Descartes to treat time as if it were another dimension of space. It gives a description of the whole history of the universe as a mathematical object, which, as we noted in chapter 1, is timeless. If you believe that it corresponds to what is objectively real in nature, you're asserting that the universe is fundamentally timeless. This block-universe picture is the second step in the expulsion of time implied by Einstein's theory of special relativity.

The block universe marries space and time. It can be pictured as a kind of spacetime, with three dimensions for space and a fourth for time (see Figure 10). An event taking place at a moment of time is represented as a point in spacetime, and the history of a particle is traced by a curve in spacetime called its world line. Thus, time has been completely subsumed by geometry; we say that time has been spatialized or geometricized. The physical laws are represented geometrically; for example, the world lines of free particles are straight lines in spacetime. If a particle is a photon, we represent it as moving at a 45-degree angle (which corresponds to measuring space in units of time, as we do when we speak of light-years). Any ordinary particle must travel slower than the photon, the bearer of light, hence its world line will be at a steeper angle.

This elegant geometrical representation of special relativity was in-

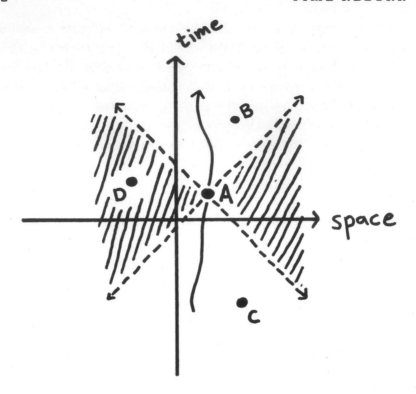

- B is in the future of A.
- C is in the past of A.
- D is causally disconnected from A.

Figure 10: The block-universe picture of spacetime. A spacetime with one spatial dimension and one time dimension. We choose the units of time and space so that light rays travel at 45-degree angles. The causal structure is then indicated geometrically; two events can be causally related if they can be connected by a line at a 45-degree angle or steeper. We also see the world line of a particle running from the past to the future through the event A. Also drawn are two light rays passing through A. The shaded regions contain events that are causally unrelated to A.

vented in 1909 by Hermann Minkowski, who had been one of Einstein's mathematics teachers. In it, every physical fact of motion implied by special relativity is represented as a theorem about the geometry of spacetime. Minkowski's invention of what we now call

Minkowksi spacetime was a decisive step to the elimination of time, because it persuasively established that all talk of motion in time could be translated into mathematical theorems about a timeless geometry. As Hermann Weyl, one of the great mathematicians of the 20th century put it: "The objective world simply is, it does not happen. Only to the gaze of my consciousness, crawling upward along the world line of my body, does a section of the world come to life as a fleeting image in space which continuously changes in time."[7]

To illustrate the power of the block-universe picture, here's a little argument some philosophers have given in support of it. The argument depends only on the relativity of simultaneity. Let's begin by agreeing that the present is real. We may not be so sure that the future or the past are real—indeed, the point of this argument is to find out how real they are—but we have no doubt that the present is real. The present consists of many events, none of which is more real than another. We don't know whether two events in the future are real, but we will agree that if two events take place at the same time they're equally real, whether that time is the present, past, or future.

If we are operationalists, we have to talk about what observers see. So we assert that *two events are equally real if they're seen by some observer to be simultaneous.* We also will assume that being equally real is what is called a transitive property; that is, if A and B are equally real, and B and C are equally real, then so are A and C. The argument then exploits the fact that the present is observer-dependent in special relativity. Pick any two events in the history of the universe, one of which is a cause of the other. Let's call them A and B. Now there will always be some other event X that has the following property: There is an observer, Maria, who sees A to be simultaneous with X. And there is another observer, Freddy, who sees X to be simultaneous with B. This is illustrated in Figure 11.

To understand why X must exist, you need to know not only that simultaneity is relative but that it is as relative as possible, in the following sense: One consequence of Einstein's postulates is that if two events take place simultaneously for some observer, all other observers will judge them to be not causally related. It's also true that if two

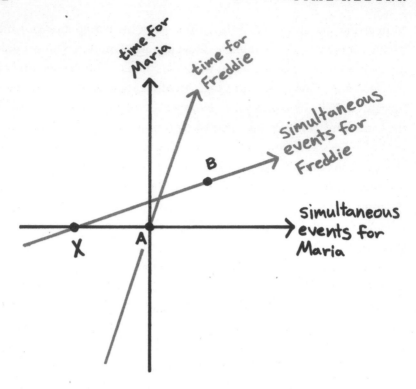

Figure 11: The argument for the block universe that follows from simultaneity.
For any two causally related events A and B, there is always an event X such
that there is an observer who sees X and A as simultaneous and an observer
who sees X and B as simultaneous.

events aren't causally related, there will be some observer who sees
them as simultaneous, thus simultaneity is as relative as it possibly
could be, while respecting causality.

If B is far in A's future, then X must be far enough away from both
so that no light signal could travel from A to X or from X to B. But the
universe that Minkwoski describes is infinite, so this is no problem.[8]

Now we can reason as follows: By the criterion I gave, A is as real as
X is. But B is also as real as X is. So A and B are equally real. But A and
B are any two causally related events in the history of the universe. So
if there is any sense in which an event in the universe is real, that reality is shared by every other event. There is thus no difference between
present, past, and future. What is real is all the events of the universe,

taken together. So we conclude that the reality of the world consists in its history taken as one. There is no reality to moments of time or their flow.

What's powerful about this block-universe argument is that to entertain it you need only believe that the present is real; the argument then forces you to believe that the future and the past are as real as the present. But if there *is* no distinction between present, past, and future — if the formation of the Earth or the birth of my great great great granddaughter are as real as the moment in which I write these words — then the present has no special claim to reality, and all that's real is the whole history of the universe.

Hilary Putnam, a leading contemporary philosopher, has reflected on this argument:

> I conclude that the problem of the reality and determinateness of future events is now solved. Moreover, it is solved by physics and not philosophy. . . . Indeed, I do not believe that there are any longer any philosophical problems about Time; there is only the physical problem of determining the exact physical geometry of the four-dimensional continuum that we inhabit.[9]

Another name for the block-universe view is eternalism. Contemporary philosophers have built up a considerable literature on its ins and outs. One question they discuss is whether the block-universe view is consistent with the way we talk about time. Ordinary people and philosophers alike use such words as "now," "future," and "past." Can these words be meaningful if reality consists of the whole history of the world as one? What do we mean when we say, "Now I'm on a train under the English Channel," if now is no more real than any other moment?

A view called compatibilism holds that there is no problem with ordinary language so long as one understands words like "now" and "tomorrow" as denoting a point of view that gives us particularly direct access to some facts about the timeless reality while making other facts hard to access. We speak easily of "here" and "there" while believing that near and far objects are equally real. So some philosophers argue that "now" and "the future" are not really very different from "here" and "there"; they all denote a certain perspective that influences what you see around you but does not affect what is real. When I use the word "now," I need not imply that now is special; I am only describing my perspective. There is always an implicit reference to which now I mean, which I assume the person I am speaking with shares.

This is well and good, but it only matters if the block universe is a correct description of nature. Other philosophers doubt that it is. John Randolph Lucas writes: "The block universe gives a deeply inadequate view of time. It fails to account for the passage of time, the pre-eminence of the present, the directedness of time and the difference between the future and the past."[10]

This is the debate to which the arguments of this book are addressed. I don't address them in the terms favored by philosophers, which are often bound up with linguistic analysis. Rather, I'm concerned with their presuppositions in physics — among them that special relativity can be applied to the whole history of the universe. But special relativity cannot be applied to the whole universe, because it doesn't contain all of physics; in particular, it doesn't contain gravity. It can be, at best, only an approximation to a theory that does contain gravity. The problem of extending relativity theory to gravity was solved by the invention of a still deeper theory, which is general relativity. This took Einstein ten years of hard work.

However, the philosophically interesting features of special relativity do extend to Einstein's theory of general relativity. The relativity of simultaneity remains true — and, indeed, is extended. So the philosophical argument I just outlined still holds and leads to the same conclusion: that the only reality is the whole history of the universe taken as one.

It also remains true in general relativity that all the information that's observer-independent is captured in causal structure and proper time. If the history of the whole universe is represented in general relativity, the result remains the block-universe picture.

◈

"Usual" Time

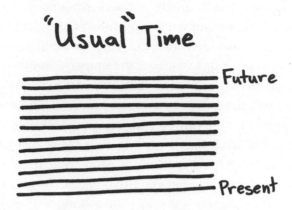

Future

Present

Time in General Relativity

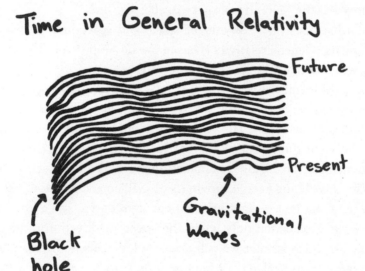

Future

Present

Black hole

Gravitational Waves

Figure 12: We contrast the usual notion of time with the more arbitrary notion in general relativity. Normally we think that time runs at the same rate everywhere, so surfaces of equal time are evenly spaced, as shown in the top image. In general relativity, time can be measured at each point by a different clock, each running arbitrarily fast compared to the others, as long as surfaces of equal time are not causally related to each other. We call this the freedom for time to be "many-fingered," as illustrated in the bottom image.

General relativity not only preserves the features of special relativity arguing that time is unreal but also introduces new ones to the same effect. First, there are many more ways of dividing spacetime up into space and time (see Figure 12). You can define time according to a network of clocks spread across the universe, but the clocks can be funky—that is, they can run at different rates in different places, and each can speed up and slow down. We describe this by saying that in general relativity time can be many-fingered. Second, the geometry of space and of spacetime is no longer simple or regular. It becomes general: Think of *any* curved surface, as opposed to simply a plane or a sphere. And geometry becomes dynamical. Waves, which we call gravitational waves, travel through the geometry of spacetime. Black holes can form, move, orbit each other. A configuration of the world is no longer given by particles positioned in space; the configuration now involves the geometry of space itself.

But what does the geometry of space and spacetime have to do with gravity? General relativity is based on the simplest of all scientific ideas, which is that falling is a natural state.

The great revolutions in physics can be marked by changes in what is considered natural motion, where by "natural" we mean a motion needing no explanation. For Aristotle, natural motion was being at rest relative to the center of the Earth. Any motion other than that was unnatural and required explanation, such as a force acting on a body to move it and keep it moving. For Galileo and Newton, natural movement was in a straight line at a constant speed, and a force was invoked as explanation only when the speed or direction of motion changed—which is what we call acceleration. This is why you don't feel motion in an airplane or a train as long as they're moving without accelerating.

You might ask, If all motion is relative, doesn't it matter whom the airplane or train is accelerating with respect to? It does, and the answer is, Other observers who are also not accelerating. Isn't this circular? It is not if we add that there is a large class of observers who feel no effect of their motion, and these have in common that they are all moving at a constant speed and direction with respect to one another. These special observers are called *inertial observers*, and Newton's laws are defined with respect to them. Newton's first law then asserts that particles which are free (in the sense that there are no forces imposed on them) travel, with respect to inertial observers, with a constant speed and direction.

This is, by the way, why it matters whether the sun or the Earth is moving. Earth's direction of motion is continually changing — with respect to any of the inertial observers — as it orbits the sun. This is an acceleration; it requires an explanation, which is provided by the gravitational influence of the sun.

For Newton, gravity was a force, like other forces. But Einstein realized that there's something special about motion impelled by gravity, which is that all bodies fall with the same acceleration regardless of their mass or any other properties. This is a consequence of Newton's laws. The acceleration of a body is inversely proportional to its mass, but Newton posits that gravity pulls on a body with a force proportional to its mass. The effects of the masses cancel, so accelerations caused by gravity do not depend on a body's mass, and all bodies fall with the same acceleration.

Einstein captured the naturalness of falling in the most beautiful principle in all of his work — and in all of physics — which he called the *equivalence principle*. This asserts that when you fall, you can't feel your motion. The experience of someone in a falling elevator is the same as that of someone free-floating in space. What we experience when we refer to gravity is the fact that we are *not* falling. The force we feel when we sit or stand is not gravity pulling us down, it's the floor or the chair pushing up on us and preventing us from falling. While I'm sitting at my desk, I am actually moving unnaturally.

This is why Einstein was a genius of the first order. Not because of the mathematical complexities of the eventual realization of general relativity — these are details that most students of mathematics and physics easily master — but because of how he succeeded in changing our perspective about one of the simplest aspects of our experience.

Before Einstein, we thought that what we felt all day, every day, was gravity pulling us down. Einstein realized we were mistaken. What we feel is the floor pushing up on us.

Einstein took this simplest and most physical of ideas and, with the help of a mathematician friend named Marcel Grossmann, turned it into a hypothesis about the geometry of the world. This hypothesis was based on a play on the simplest notion of geometry, that of a straight line.

A straight line is defined in high school geometry as the path that takes the shortest distance between two points. This definition applies to a plane but can be extended to curved surfaces. Think of a sphere, like the surface of the Earth. You might think there are no longer any straight lines, because the surface is curved, but this is not the case if what we mean by a straight line is the path that takes the shortest distance between two points. We call curves satisfying this definition *geodesics.* When the space is a plane, the geodesics are the straight lines; when the space is a sphere, the geodesics are segments of great circles, and they are the routes airplanes take as the shortest trip between cities.[11]

If the paths of objects falling in a gravitational field are the natural motions, they should generalize the straight lines along which, according to Newton, objects with no forces on them naturally move. But now we have a choice, because just as free particles move along straight lines in space, they move along straight lines in Minkowski spacetime. So do we want to represent gravity by curving space or by curving spacetime?

From the block-universe perspective, the answer is clear: It must be spacetime that's curved. Because of the relativity of simultaneity, different observers differ as to which events are simultaneous. There is no simple, objective, observer-independent way to describe how space is curved.

When Einstein chose to realize the equivalence principle by curving spacetime, his idea was that the curvature would transmit the influence of gravity in such a way that objects falling in a gravitational field would move along geodesics. Free-falling bodies would fall toward Earth not because they feel a force but because spacetime has been curved so that the geodesics arc toward Earth's center. Planets orbit the sun not because the sun exerts a force on them but because its enormous mass curves the geometry of spacetime so that the geodesics curve around it.

This was how Einstein explained gravity as an aspect of the geometry of spacetime. Geometry acts on matter by guiding it along geodesics. But what is wonderful about Einstein's theory of general relativity is that this action is reciprocated. Einstein posited that mass causes geometry to curve so that geodesics accelerate toward massive bodies. To implement this idea, he proposed equations that direct spacetime to curve in just the right way as to mimic the effects of gravity.

These equations have many consequences that have been confirmed by observation to high precision. They cause the universe as a whole to expand. They predict that the orbits of planets around the sun and the moon around the Earth differ very slightly from those predicted by Newtonian physics, and these effects have been observed. They cause the spacetime around extremely compact bodies to curve so much that no light can escape; these are black holes, and there are tremendously massive ones — as massive as many millions of stars — at the center of most galaxies.

Perhaps the most remarkable consequence of the equations of general relativity is that the geometry of spacetime is distorted by the passage of waves through it. These are analogous to the distortions in the surface of a pond; it is the geometry of space that oscillates as the waves travel through. These gravitational waves are caused by rapid changes in the motions of very massive bodies, such as two neutron stars orbiting each other, and they carry images of these violent events across the universe. A great effort is currently under way to detect these images, which will open up a new window into astronomy that will enable us to see inside of collapsing supernovas and all the way back to the first moments of the Big Bang — and possibly before.

The effects of gravity waves have been seen indirectly. When two neutron stars rotate rapidly around each other, the gravitational waves they produce take energy away, causing them to spiral toward each

other. This spiraling-in has been observed and found to agree to a high
degree of precision with predictions of general relativity.

◈

With his invention of general relativity, Einstein instigated a radical
transformation in the conception of space and time.

In Newton's physics, the geometry of space is fixed once and for all.
Space is assumed to have the geometry of three-dimensional Euclidean space. There is then something worryingly asymmetric about the
relationship between space and matter in Newtonian physics: Space
appears to tell matter how to move, but space itself never changes.
There's no reciprocation. Space is never influenced by the motion of
matter or even its presence. Space, it seems, would be exactly the same
were there no matter in it at all.

This asymmetry is corrected in general relativity, in which space
becomes dynamical. Matter influences the changes in geometry just
as geometry influences the motion of matter. Geometry becomes fully
an aspect of physics, just like the electromagnetic field. The Einstein
equations that spell out the dynamics of spacetime are then like other
hypotheses: They probe the properties of physical phenomena and
their relationships with one another.

Were the geometry of spacetime fixed for all time, we would say
that space and time are absolute: Only the details differ from Newton's
conception of the properties of space and time as timeless and fixed.
That geometry is dynamical and influenced by the distribution of matter realizes Leibniz's idea that space and time are purely relational.

In his formulation of a relational theory of space and time, Einstein
was guided by Ernst Mach, who introduced a principle we call *Mach's
principle.* This says that only relative motion should matter, so that if
we get dizzy when we spin, it must be because we are spinning with respect to the distant galaxies. The claim that the effect is one of purely
relative motion implies that we would feel equally dizzy were we to
stay still and the whole universe spin around us.

But while general relativity is radical in these respects, it is conservative in another, which is that it fits neatly into the Newtonian paradigm. There is a space of possible configurations of the geometry and the matter together. Given the initial conditions, Einstein's equations determine the whole future geometry of a particular spacetime and everything it contains, including matter and radiation.

And the whole history of the world is, in general relativity, still represented by a mathematical object. The spacetime of general relativity corresponds to a mathematical object much more complex than the three-dimensional Euclidean space of Newton's theory. But seen as a block universe, it is timeless and pristine, with no distinction of future from past and no role for or sign of our awareness of the present.

❖

General relativity had one more blow to deliver to time's fundamental role in physics. Implicit in the idea that time is real and fundamental is that time cannot have a beginning. For if time has a beginning, then that origin of time must be explicable in terms of something that is not time. And if time is explicable in terms of something timeless, then time is not fundamental and whatever it is that time emerges from is more fundamental. But in any plausible model of a universe described by the equations of general relativity, time always has a beginning.

Within a year of the 1916 publication of his theory of general relativity, Einstein applied it to the whole universe. He did this by imagining that the universe was finite in extent but without a boundary—like a sphere. This was a profound step; for the first time, the universe could be seen as self-contained and finite. There is only so much to it, but there is no way to get outside it. "Outside the universe" has no meaning at all.

In closing the universe, Einstein had to assume that any clocks used to measure time were inside the system. He could do this because the equations of his theory had a novel feature, which is that they made sense no matter what clocks were used to measure time and what devices were used to measure space. Time and space could be measured as funkily and messily as possible and the equations still worked. So the theory was no longer tied to measurements made by special clocks running outside the system.[12] By removing the need for a clock outside the system, general relativity goes some distance toward a relational theory of physics. But it still is based on the Newtonian paradigm, as it can be formulated in terms of timeless laws acting on a timeless configuration space.

At first, Einstein sought to model a universe that was not only finite in space but eternal and unchanging in time. As original in his thinking as any scientist we know of, Einstein's imagination failed him here, for it does not seem to have occurred to him to conceive of the universe as anything but static and eternal. But there was a problem, which is that the gravitational force is universally attractive and always acts to bring things together. This means that gravity acts on the whole universe to cause it to contract. If the universe is expanding, gravity will slow the expansion. If it is neither expanding nor contracting, gravity will act to begin a contraction. Einstein thus might have predicted that the universe must be changing in time, either expanding or contracting. Instead, he changed his theory in an attempt to keep the universe static and so made a different and unintentional discovery — one that was not confirmed by experiment until recently.

Einstein modified his equations by adding a term that counters gravity by causing the universe to expand. This modification came with a new constant of nature, representing an energy density of empty space. Einstein called it the *cosmological constant.* There is good evidence for it in the recently observed acceleration of the expanding universe. A more general name for the cause of the accelerated expansion is *dark energy,* but if its density is constant in space and time it can be described by Einstein's cosmological constant. So far the observations are consistent with this, but several cosmological scenarios require the dark energy to eventually vary.

I don't think Einstein ever imagined that this constant would one day be measured, but it has been. It has an incredibly tiny value — and correspondingly enormous consequences. Even though it's tiny, its effects add up across the universe. Thus there are two opposing forces acting on the universe. Gravity from all the matter causes contraction, while the cosmological constant accelerates expansion.

Einstein proposed a static universe, in which these forces were exactly balanced. But there was a problem with this, too — which is that the balance was unstable. Tweak the universe just a bit and one tendency would win, so the universe should either expand forever or contract. The universe is full of moving stars, black holes, and gravitational waves, and they would provide enough tweaking to guarantee that it couldn't stay balanced for long.

The astounding conclusion is that the universe must have a history. It can expand and it can contract, but it cannot stay the same. In the 1920s, several astronomers and physicists discovered solutions to the equations of general relativity — solutions that describe expanding universes. This was fortunate, because by 1927 the astronomer Edwin Hubble had discovered evidence that the universe was expanding — which implied that it must have had a beginning. And, indeed, every one of these new solutions that was not unstable had a first moment of time.

Those solutions are associated with the names of Alexander Friedmann, H. P. Robertson, Arthur Walker, and Georges Lemaître and are called FRWL universes. They are very simple models, in that they assume that the universe is the same everywhere in space: That is, the same density of matter and radiation obtains throughout. At the first moment of time in a FRWL universe, the density of matter and radiation and the strength of the gravitational field become infinite and constitute an *initial singularity*. At that point, general relativity stops working, because the equations no longer describe the evolution of the future out of the present. The infinite quantities cause them to break down.

The response of most physicists was that the equations broke down because the models being studied were too simple. They claimed that if you put in more details — so that the universe could have local features, like stars, galaxies, and gravitational waves — the singularity would be eliminated and you could continue to extrapolate time back beyond that point. This hypothesis was hard to confirm, because in the era before supercomputers it was impossible to fully study general solutions to the equations of Einstein's theory. So for several decades the hypothesis survived simply because it was difficult to test. But it turned out to be wrong. In the 1960s, Stephen Hawking and Roger Penrose proved a theorem stating that there are singularities in all solutions to the equations of general relativity that might describe our universe.

If general relativity is a true description of our universe, it's hard to escape the conclusion that time cannot be fundamental. Otherwise we have some embarrassing questions to answer—such as, What happened before time started? And what started the universe? Even more puzzling are questions about timeless laws: If laws are timeless, then what were they doing before there was a universe for them to govern? Clearly the answer is that there *was* no time before the universe, which means that the laws must be a deeper aspect of the world than time.

In some of these solutions, time, once started, goes on forever, as the universe forever expands and dilutes. But in other solutions, the universe reaches a maximal expansion and then collapses to a Big Crunch, in which many observable quantities become again infinite; these latter solutions describe universes in which time, too, has an end. Time starting and stopping is no problem for the block-universe picture, within which what's real is the history of the universe taken as a timeless whole. That reality is not compromised if it involves a world in which time begins or ends. Instead, the discovery that time begins in solutions to general relativity that describe a whole universe strengthens the block-universe picture as it weakens any claim that time is more fundamental than law.

We have come a long way in the story of the expulsion of time from the physicists' conception of nature. We began, as Galileo and Descartes began, by capturing motion and freezing time through their method of graphing, in which time is represented as if it were another dimension of space. In relativity theory, these pictures of motion laid out in time become spacetime, a timeless picture of the history of the universe in which there is nothing real about the present moment. The relativity of simultaneity tells us that we cannot go back to separate time from space. We can only go ahead to the block-universe picture, in which the history of the universe is presented as a timeless whole. With special and general relativity well confirmed by experiment, we physicists have indeed much reason to embrace a timeless picture of reality.

7

Quantum Cosmology and the End of Time

OVER CHRISTMAS BREAK at the end of my first semester at Hampshire College, I came down to New York City to stay in my cousin's Greenwich Village apartment. In the morning, I took the subway uptown to my first physics conference, grandly titled the 6th Texas Symposium on Relativistic Astrophysics, which was held at a fancy hotel in midtown Manhattan. I wasn't invited and I don't think I registered, but my physics professor, Herb Bernstein, had suggested I drop in. I didn't know a soul there, but somehow I met Kip Thorne, of Caltech, who told me that to learn general relativity well I should study from the textbook he had just written with Charles Misner and John Archibald Wheeler.[1] I met Lane Hughston, a young American mathematician studying at Oxford, who took an hour to explain the revolutionary new twistor theory to me and then introduced me to its inventor, Roger Penrose.

In one session, I had taken a seat on the aisle when a man in a wheelchair motored in next to me. Stephen Hawking was already famous for his work on general relativity, and this was a year before his astounding discovery that black holes are hot. A tall bearded man with elegant manners stopped to chat with him and then was called to the stage. This was Bryce DeWitt. I don't recall what he talked about, but I had heard of him and his equations describing quantum universes. I didn't have the courage to speak to either of them, and I certainly never imagined that when I finished my PhD seven years later, these two giants of modern physics would invite me to work with them.

Bryce DeWitt, John Wheeler, Charles Misner, and Stephen Hawking were all pioneers who were then in the midst of creating a new subject: quantum cosmology. The marriage of general relativity with quantum theory they invented is the summit of our climb up to the timeless world of contemporary physics. In the quantum universe they described, time is not just redundant, it disappears completely. The quantum cosmos doesn't evolve or change, it doesn't expand or contract, it simply is.

This subject, it must be stressed, is a highly speculative and poetic domain of theoretical physics with as yet no solid connection to observation. The conclusions you can draw from it lack the authority of the picture of nature given by relativity theory, which has triumphed experimentally over and over again and continues to surprise us with the accuracy of its predictions.

We begin with quantum mechanics, which is a triumph of the method of doing physics in a box. I'll need to explain just a few basic things about how the universe's subsystems are modeled in quantum mechanics, to pave the way for a two-step extrapolation of our current physics. First, we need to unify quantum mechanics with general relativity to get a quantum theory of gravity. There are different approaches to this unification, and as yet no experiment to decide among them, but enough is known about how such a theory might be formulated to let us proceed to the second step, which is the inclusion of the whole universe into quantum theory.

We will see that the result is a timeless picture of nature.

Quantum mechanics offers a very successful description of microscopic systems, such as atoms and molecules. But it is baffling. As a result of people's attempts to make sense of it, several radically different ways of talking about it have been invented. These differ in their implications concerning time and whether quantum theory can be applied to the whole universe — both prime topics for our discussion here.[2]

In my view, the best way to explain quantum mechanics is to start by talking about what science is for. Many of us think the purpose of science is to describe how nature really is — to give a picture of the world that we can believe would be true, even were we not here to see it. If you think of science that way, you'll be disappointed by quantum mechanics, because it gives no picture of what is going on in an individual experiment.

Niels Bohr, one of the founders of quantum theory, argued that those who were disappointed in this way had the wrong idea of what science is for. The problem is not the theory but what we expect a theory to do for us. Bohr proclaimed that the purpose of a scientific theory is not to describe nature but to give us rules for manipulating objects in the world and a language we can use to communicate the results.

The language of quantum theory assumes an active intervention in nature, for it speaks about how an experimenter interrogates a microscopic system. She can prepare the system so that it's isolated and ready to be studied. She can transform it by subjecting it to various outside influences. And then she can measure it by introducing devices that read out the answers to questions she might like to ask of the system. The mathematical language of quantum mechanics represents each of the steps in the process of preparation, transformation, and measurement. Because of the emphasis on what we do to a quantum system, this can be called an operational approach to quantum physics.

The central mathematical object in the quantum description of a system is called a *quantum state*. It contains all the information an observer can know about a quantum system as a result of preparing and measuring it. This information is limited, and in most cases it does not suffice to predict exactly where the particles that make up the system are. Instead, a quantum state gives *probabilities* for where we might find the particles if we were to measure their positions.

Consider an atom, consisting of a nucleus with some electrons around it. The most precise description you could give of the atom would be to say where each electron is. Each arrangement of the electrons is a configuration. Quantum mechanics' best description is instead to give a probability for each possible configuration in which the electrons might be found.[3]

How do you check the predictions of a theory, when those predictions are only probabilistic? Think of the prediction that a tossed coin lands heads-up 50 percent of the time. To check this, you cannot just toss a coin once; the result will be either heads or tails, and either is consistent with the prediction that each occurs half the time. You need to toss a coin many times and record what proportion of the tosses end up heads. As you toss more and more times, the proportion of heads will tend toward 50 percent. It's the same with the probabilistic predictions of quantum mechanics: To confirm them, you need to do an experiment many times.[4] Measuring a single quantum system is like tossing a coin just once: Whatever random result you get will be consistent with almost any prediction of the theory.

This method makes sense only when applied to a small, isolated system, such as a hydrogen atom. To check the predictions, we require a large number of identical copies of the system; if we have just one, we can't check the predictions — because they're probabilistic! We must also be able to manipulate this collection of systems, preparing them initially in the quantum state we're interested in and then measuring something about them. But if we have many copies of a system in the world, then each copy must be just a small part of what exists. Among the things that aren't part of the system are the instruments and coordinate axes we use to measure the system's configurations.

So the application of quantum mechanics appears to be limited to isolated systems. It's an extension of the Newtonian paradigm — of doing physics in a box. To see how firmly the method of quantum mechanics is based on the study of isolated systems, let's look at how change in time is described.

The laws of Newtonian physics are deterministic, which means that the theory gives definite predictions for how a system evolves in time.

Likewise, a law of quantum mechanics tells us how a system's quantum state evolves in time. This law is also deterministic, in the sense that given an initial quantum state, you can predict exactly what the quantum state will be at a later time.

The law of evolution of quantum states is called the *Schrödinger equation*. It works just like Newton's laws, but it tells how states, rather than the positions of particles, change in time. If you input an initial quantum state, the Schrödinger equation will tell you what the quantum state will be at any later time.

As in the case of Newtonian physics, the clock must be outside the system, along with the observers and their measuring instruments.

But although the evolution of the quantum state is deterministic, the implications for the precise configurations of the atoms are only probabilistic — because the connection between the quantum state and the configurations is itself probabilistic.

The requirement that the clock that measures time in quantum mechanics must be outside the system has stark consequences when we attempt to apply quantum theory to the universe as a whole. By definition, nothing can be outside the universe, not even a clock. So how does the quantum state of the universe change with respect to a clock outside the universe? Since there is no such clock, the only answer can be that it doesn't change with respect to an outside clock. As a result, the quantum state of the universe, when viewed from a mythical standpoint outside the universe, appears frozen in time.

This is an admittedly slick verbal argument that seems as if it could easily result in misleading conclusions. But in this case the math holds up, giving us the same result as when we apply Schrödinger's equation to the quantum state of the universe: The state doesn't change in time.

In quantum theory, change in time is connected with energy. This is a consequence of a basic feature of quantum physics called the wave-particle duality.

Newton understood light to be made up of particles. Later, the phenomena of diffraction and interference were studied, and to explain these it was hypothesized that light is a wave. In 1905, Einstein resolved the dilemma as to the nature of light by proposing that light has aspects of waves and aspects of particles. Almost two decades later, Louis de Broglie proposed that this duality of wave and particle is universal: Everything that moves has some aspects of a wave and some aspects of a particle.[5]

This may seem mysterious. It is surely impossible to visualize something that is both a wave and a particle. Exactly! As I've noted, quantum mechanics describes phenomena that cannot be visualized. We can manipulate quantum particles in experiments and talk about how they respond to being measured, but we cannot visualize what goes on in the absence of our manipulation of nature.

One wave aspect of light is its frequency, the number of times it oscillates per second. A particle aspect of light is its energy; each particle of light carries a certain amount of energy. In quantum mechanics the energy in the particle picture is always proportional to the frequency in the wave picture.[5]

Armed with this understanding of the wave-particle duality, let's return to the quantum state of the universe. Because there is no clock outside the universe, the quantum state of the universe cannot change in time. So its frequency of oscillation must be zero — if it is frozen, it cannot oscillate. But since frequency is proportional to energy, this means the energy of the universe must be zero.

There is a negative amount of energy trapped in any system held together by gravity. Consider the solar system. If you wanted to pull Venus out of its orbit around the sun, and remove it from the solar system, it would take energy. Since you have to add energy to bring Venus into a state where it has no energy, while it remains trapped in its orbit Venus has negative energy. This negative energy is called the gravitational potential energy.

The universe can have zero total energy if the total gravitational potential energy holding all its parts together exactly cancels all the positive energy in the universe, expressed in the masses and motions of all its matter.

Having zero energy and frequency, the quantum state of the universe is frozen. The quantum universe does not expand or contract. There are no gravitational waves moving through it. There is no galaxy formation, no planets orbiting stars. The quantum universe simply is.[6]

These consequences of applying quantum mechanics to the whole universe were discovered in the middle 1960s by the pioneers of quantum gravity: DeWitt, Wheeler, and Peter Bergmann. The modification of the Schrödinger equation we alluded to — with the condition that the quantum state is frozen — is named after two of them and is called the *Wheeler-DeWitt equation*. Pretty quickly, they noticed the disappearance of time and people began arguing about the implications. We're still arguing. Every few years, someone hosts a conference devoted to the Problem of Time in Quantum Cosmology. Human ingenuity is limitless, so a wide range of responses and proposals have been offered.

The frozen state of the universe is not the only thing that goes wrong when we attempt to apply quantum theory to cosmology.[7]

There's only one universe, so you can't construct a population of systems in identical quantum states and compare the measurements made on them with the probabilities predicted by quantum mechanics. Right away, the scope for comparing theory to experiment or observation narrows dramatically.

It's even worse than this, because you cannot prepare the universe in an initial quantum state, let alone study the consequences of different choices of initial state. The universe happened just once and had whatever initial state it had. We weren't there to choose its initial state, and even if we had been, we couldn't have manipulated the universe, because we would have been part of it. The very idea of preparing the universe in an initial state imagines for us a god-like status of existing outside the universe.

The tragedies of quantum cosmology amount to a formidable list: We cannot prepare the initial state of a quantum universe and we cannot act on it from outside the universe to transform it. We don't have access to a collection of universes to give meaning to the probabilities the quantum formalism outputs. On top of that, there's no place outside the universe to put our measuring instruments. So there's no notion of measuring change by a clock external to the quantum system under study.

From an operational point of view, applying quantum mechanics to the universe was crazy to begin with. It failed because we applied it in a context in which none of the operational definitions that define the theory make sense. All this is payback for committing the fallacy of extending a method well adapted to small parts of the universe to the whole of it.

The problem is even a bit harder than noted here, because the choice of a time coordinate in general relativity is, as we saw, completely arbitrary. So you have to ask, "If there *were* a clock outside the universe, which notion of time within the universe would it correspond to?" And "If there *were* a quantum state that was oscillating, which clocks in the universe would pick that up as a regular oscillation?" The answer is, "Every possible notion of time and every possible clock." As a result, there is not one Wheeler-DeWitt equation but an infinite number of them. They assert that the frequency with which the quantum state oscillates is zero for every possible notion of time and every possible clock within the universe. For every possible clock, carried by every possible observer, nothing happens in the quantum universe.

All this remained academic for two decades, because no one could actually solve the Wheeler-DeWitt equations. Not until the invention of the approach to quantum gravity called loop quantum gravity was there a context in which these equations could be formulated precisely enough to be solved. This revolution was set in motion by Abhay Ashtekar's discovery of a new formulation of general relativity in 1985.[8] A few months later, I was fortunate enough to be working at the Institute for Theoretical Physics (now called the Kavli Institute for Theoretical Physics) at UC Santa Barbara with Ted Jacobson (now at the University of Maryland), and we found the first exact solutions to the Wheeler-DeWitt equations — in fact, an infinite number of them.[9] There were other equations needing to be solved to write a complete quantum state of the gravitational field, and this was accomplished two years later with Carlo Rovelli, then at the Istituto Nazionale di Fisica Nucleare, University of Rome.[10] Matters progressed rapidly, and a much larger set of solutions was discovered by Thomas Thiemann at Harvard in the early 1990s.[11] Since then, even more powerful techniques for generating solutions, based on what we now call spin-foam models, have been developed.[12] These results increase the urgency to resolve the problem of time in a timeless universe, in order to give physical meaning to all these mathematical solutions to a theory of quantum gravity.

The crux of the issue is whether time can be said to "emerge" from the timeless universe so that the theory is not in blatant conflict with the aspects of time we see acting in the world. Some of my colleagues suggest that time is part of an approximate description of the universe — a description that is useful on large scales but dissolves when we look too closely. Temperature is like this: Macroscopic bodies have temperatures, but single particles don't, because the temperature of a body is the average of the energies of the atoms that make it up. Some physicists have proposed that time, like temperature, is meaningful only in the macro world but not relevant at the Planck scale. Other approaches aim to discover time in correlations between different subsystems of the universe.

I have spent countless hours pondering these approaches to how time could emerge from a timeless universe, and I remain unconvinced that any of them work. In some cases the reasons are technical and would not be useful to describe here. The deeper reasons for my skepticism regarding quantum cosmology will be our focus in Part II.

My friends on the other side of this debate argue that the assumptions leading to the Wheeler-DeWitt equations involve only the principles of quantum mechanics and general relativity taken together. Given that these principles are each, in their individual domains, well confirmed experimentally, it is wise to first take all their implications seriously and try to understand and develop them.

When I was one of his postdocs, Bryce DeWitt used to urge us not to impose our metaphysical prejudices on a theory but to let the theory dictate its own interpretation. I can still hear him exhorting us, in his gentle voice, to "let the theory speak."

The best thought-through approach to making sense of the quantum cosmology framed by the Wheeler-DeWitt equations has been proposed by the British physicist, philosopher, and historian of science Julian Barbour, who described it in his 1999 book, *The End of Time.* Barbour's idea is radical but not difficult to explain in words. He asserts that all that exists, fundamentally, is a vast collection of frozen moments. Each moment has the form of a configuration of the universe. Each configuration exists — and is experienced by any beings caught in that configuration — as a moment of time. Barbour calls the collection of all the moments "the heap of moments." The moments in the heap don't follow each other, one after the other. There's no order to them. They simply are. In Barbour's metaphysical picture, nothing at all exists except these pure moments of time.

You may object, "But I experience time passing." Barbour says you don't. All any of us experience, he insists, is moments — snapshots of experience. Snap your fingers — that was one snapshot, one moment in the heap. Snap your fingers again — that's another moment. You have the impression that the second followed the first, but that's an illusion. You think so because in the second moment you have a memory of the first moment. But that memory is not an *experience* of time passing (which Barbour says never happens); it's just that the memory of the first moment is part of the experience of the second moment. All we experience — and all that's real, according to Barbour — are the individual moments in the heap.

However, there's one bit of structure in the heap. Moments can be represented more than once. You can then speak of the *relative frequency* of moments; one moment may be present a billion times more than another moment.

These relative frequencies of moments are what the probabilities given by the quantum state refer to. Two configurations have a relative probability to appear in the heap, given by their relative probability in the quantum state.

That's all there is. There's one quantum universe, described by one quantum state. That universe consists of a very large collection of moments. Some are more common than others. Some, indeed, are vastly more common than others.

Some of the common configurations in the heap are boring. They describe a moment of time in a universe filled with a gas of photons, or a gas of hydrogen atoms. Barbour asserts that in the actual quantum state of the universe, most of these boring configurations have small volume. Hence he predicts a correlation between being small and being boring. If we assumed the existence of time, we would say that the universe *was* boring *when* it was small. Barbour says it is enough to say that being small and being boring are highly correlated properties of moments in the heap.

Other configurations in the heap are interesting — full of complexity, with living beings like ourselves living on planets orbiting stars in galaxies, which are themselves arranged in sheets and clusters. Barbour asserts that in the right quantum state, the property of being full of complexity and life correlates with a large volume. Thus many, and perhaps most, of the configurations in the heap with large volumes will have living beings in them.

Furthermore, Barbour asserts that in the right quantum state, the most common configurations have structures that refer implicitly to other moments. These references are what Barbour calls "time capsules." They are memories, books, artifacts, fossils, DNA, and so on. They tell a story open to interpretation in terms of a sequence of moments in which things happened that built on each other, leading to complexity. That is, the time capsules support the illusion that time is passing.

According to Barbour's theory, causality, too, is an illusion. Nothing can be the cause of anything, because in actuality nothing is happening in the universe. There's just a vast pile of moments, some of which are experienced by beings like ourselves. In reality, each experience of each moment is only that — unconnected to the rest. There are moments, but there is no ordering of them, no passage of time.

But the Wheeler-DeWitt equations do allow approximate notions of order and causality to emerge, so that there are correlations among the most common moments — correlations that make it appear as if there were a succession of moments in time, between which causal processes could operate. To a high degree of approximation, the story of a succession of moments can be helpful for explaining the structures that occur in the moments. But it is not a fundamental story, and when looked at closely enough, there is no order and no causality, just a pile of moments.

Barbour's theory has its elegant features. It neatly solves the problem of what the probabilities in quantum cosmology can refer to. There's only one universe, but it contains many moments. The quantum probabilities are real relative-frequency probabilities for moments to exist as part of reality. To the extent that Barbour's scheme works in detail, it explains how the impression emerges that the world has a history during which causal processes contribute to building complex structures. This proposal also explains the apparent directionality of time: There is a preferred direction in configuration space, which is away from the configurations of small volume and toward larger volumes. When time emerges, increasing volume correlates well with increasing time, so this explains why the universe appears to have an arrow of time.

Barbour's version of timeless quantum cosmology offers palpable consolation for our mortality. I can feel it. I wish I could believe it. You experience yourself in a collection of moments. According to Barbour, that's all there is. Those moments always are, eternally. The past is not lost. Past, present, and future are with us, always. Your experience may figure in a finite set of moments, but those moments never go away or cease. So nothing comes to an end when you come to your last day. It's just that *now* you are experiencing a moment that has all the memories you will ever have. But nothing ceases, because nothing ever started. The fear of death is based on an illusion, which in turn is based on an intellectual mistake. There's no flow of time running out, because there's no flow of time. There are just, and always are, and always will be, the moments of your life.

I won't speculate about what Einstein would have thought of Julian Barbour's timeless quantum cosmology. But there's evidence that he found much satisfaction and solace in the disappearance of time in the block-universe picture. From his teenage years, he had sought to transcend the messy human world through the contemplation of the timeless laws of nature. In a letter of consolation to the widow of his friend Michele Besso, he wrote: "Now he has departed from this strange world a little ahead of me. That means nothing. People like us, who believe in physics, know that the distinction between past, present, and future is only a stubbornly persistent illusion."

PART II

LIGHT: TIME REBORN

INTERLUDE

Einstein's Discontent

THE BLOCK UNIVERSE of Einstein's theories of relativity was the definitive step in the expulsion of time from physics. But Einstein himself was ambivalent about the disappearance of time from the conception of nature he had done so much to build. We saw how he found consolation in the block-universe picture of the timeless cosmos — yet it appears that Einstein was not content with its implications. We know this from the *Intellectual Autobiography* of the Viennese philosopher Rudolf Carnap, who reported a conversation with Einstein on time:

> Once Einstein said that the problem of the Now worried him seriously. He explained that the experience of the Now means something special for man, something essentially different from the past and the future, but that this important difference does not and cannot occur within physics. That this experience cannot be grasped by science seemed to him a matter of painful but inevitable resignation.

If Einstein was reflective, Carnap had no doubt where he himself stood:

> I remarked that all that occurs objectively can be described in science; on the one hand the temporal sequence of events is described in physics; and, on the other hand, the peculiarities of man's experiences with respect to time, including his different attitude towards past, present, and future, can be described and (in principle) explained in psychology.

I cannot imagine what Carnap was thinking. I know of no way in which the sciences of psychology or biology could account for our experience of time in a timeless world.[1] Einstein was not satisfied by Carnap's reply either, according to Carnap, who writes, "But Einstein thought that these scientific descriptions cannot possibly satisfy our human needs; that there is something essential about the Now which is just outside the realm of science."[2]

Einstein's discontent comes down to a simple insight. A scientific theory, to be successful, must explain to us the observations we make of nature. Yet the most elemental observation we make is that nature is organized by time. If science must tell a story that encompasses and explains everything we observe in nature, shouldn't that include our experience of the world as a flow of moments? Isn't the most basic fact about how experience is structured a part of nature that a fundamental theory of physics should incorporate?

Everything we experience, every thought, impression, action, intention, is part of a moment. The world is presented to us as a series of moments. We have no choice about this. No choice about which moment we inhabit now, no choice about whether to go forward or back in time. No choice to jump ahead. No choice about the rate of flow of the moments. In this way, time is completely unlike space. One might object by saying that all events also take place in a particular location. But we have a choice about where we move in space. This is not a small distinction; it shapes the whole of our experience.

Einstein and Carnap agree about one thing: that experiencing nature as a series of present moments is not part of the physicists' conception of nature. The future of physics — and, one might add, the physics of the future — comes down to a simple choice. Do we agree with Carnap that the present moment has no place in science, or do we follow the instinct of the greatest scientific intuition of the 20th century and try to find a way to a new science in which Einstein's "painful resignation" will not be necessary?

For Einstein, the present moment is real and should somehow be part of an objective description of reality. He believes (as Carnap puts it) that *"there is something essential about the Now which is just outside the realm of science."*

At least sixty years have passed since that conversation took place. We have learned a great deal about physics and cosmology since then. Enough to bring the Now, at last, into the physicists' description of nature. In this second part of the book, I will explain why our current knowledge requires that time be reinserted as a central concept in physics.

In Part I, we traced nine steps in the expulsion of time from the physicists' conception of nature, beginning with Galileo's discoveries about falling bodies and on up to Julian Barbour's timeless quantum cosmology. Shortly we will see time reborn, but first we have to deconstruct the apparently strong arguments given in Part I.

The nine arguments fall into three classes:

Newtonian arguments (that is, arguments stemming from Newton's physics or Newton's paradigm for doing physics):

- The freezing of motion by graphing records of past observations
- The invention of the timeless configuration space
- The Newtonian paradigm
- The argument for determinism
- Time-reversibility

Einsteinian arguments, stemming from the theories of special and general relativity:

- The relativity of simultaneity
- The block-universe picture of spacetime
- The beginning of time in the Big Bang

Cosmological arguments, stemming from extending physics to the universe as a whole:

- Quantum cosmology and the end of time

These nine arguments lead to a view of nature that denies the reality of the present moment and instead speaks of nature in terms of the block universe picture in which what is real is only the entire history of the world taken as one. In this picture, time is treated as a dimension of space, so causation within time can be replaced by timeless logical inference. General relativity and Newtonian mechanics may speak of histories evolving over time, but this is time in a weak sense of mathematical order only, stripped of the coming into being of present moments. In these theories, time is not real, in the sense I defined in the preface when I asserted that all that is real is such in a moment of time. To make the contrast vivid I will refer to such theories as timeless.

Is this expulsion of time the price that must be paid for the progress of science? The next step in our journey is to reveal the flaws in these arguments.

All nine arguments labor under a common fallacy: that the Newtonian paradigm — which assumes that we can predict the future state of any system from its initial conditions and the laws acting on it — can be extended to make a theory of the universe. But, as I will shortly show, no extension of the Newtonian paradigm can yield an acceptable theory of the universe as a whole. Although it's a powerful method when applied to physics in a box, it becomes impotent when confronted with cosmological questions.

The strongest arguments for the elimination of time have come from relativity theory. In chapter 14, we will break these down. Once we have deconstructed the case for the elimination of time, we will consider what physics and cosmology gain from the hypothesis that time is real.

8

The Cosmological Fallacy

I N PART I, WE FOLLOWED the path of the mystic, seeking to
transcend our time-bound experience and discover eternal truths.
In particular, we traced the great success of physics to its use of
a method, the Newtonian paradigm. We saw that this success came
with a price: the expulsion of time from the physicists' conception of
nature.

In Part II, we will see why the price need not be paid, because the
attempt to apply the Newtonian paradigm to the universe as a whole is
an impossible task. To extend science to an understanding of the universe as a whole, we need a new theory — one in which the reality of
time is a central element.

Let's go back to the beginnings of science, to the person who has
been called the first scientist, the pre–Socratic philosopher Anaximander (610–546 BC). As described in a recent book by Carlo Rovelli,
Anaximander was the first to seek the causes of natural phenomena in
nature itself rather than in the capricious will of the gods.[1]

In those days, even the most knowledgeable human beings thought
themselves inhabitants of a universe framed by two flat media. Below

our feet was the Earth, stretching out in every direction around us. Above our heads was the sky. The entire universe, as they understood it, was organized by the presence of a special direction — down, the direction in which things fall. The basic law of nature, confirmed by all of their experience, was that things fall downward. The only exception was the sky itself and the heavenly bodies fixed there.

When they attempted to extend this successful law to the universe (Earth and sky), they encountered a paradox: If everything not fixed to the sky falls down, why doesn't the Earth itself fall? Since the tendency to fall down is universal, the Earth must have something under it, holding it up — one such proposal was that Earth rested on the back of a giant turtle. But then, what held that turtle up? Might there be an infinite array of "turtles all the way down"?

Anaximander realized that a conceptual revolution was needed to make a successful theory of the universe that avoided the *reductio ad absurdum* of an infinite tower of turtles. He proposed an idea obvious to us but shocking in its time — that "down" is not a universal direction but simply the direction toward the Earth. The right way to state the law is not that things fall down, it is that things fall toward the Earth. This would enable another revolution — the discovery that the Earth is not flat but round. Anaximander himself did not take that breathtaking step, but his redefinition of "down" freed him to see the Earth as a body afloat in space. Thus he could make the amazing suggestion that the sky extended all around the Earth — under our feet as well as over our heads.

This insight greatly simplified the cosmology of the time, because the fact that the sun, moon, and stars rise in the east and set in the west could be understood as due to a daily rotation of the sky. It was no longer necessary to make the sun anew each morning in the east and destroy it each evening after it set in the west; after sunset, the sun returned to its starting point by taking a path beneath our feet. Imagine the elation of understanding that for the first time! This removed a great source of ancient anxiety — that whatever spirit was responsible for making a new sun each morning might oversleep or desert its post. Anaximander's revolution was arguably greater than Copernicus's, because his redefinition of "down" rendered moot the need to explain what held the Earth up.

The philosophers who sought to understand what holds the Earth up were making a simple mistake — taking a law that applies locally and applying it to the whole universe. Their universe was Earth and sky, and ours is a vast cosmos filled with galaxies, but the same mistake underlies much of the confusion of current cosmological speculation. And yet nothing seems more natural, for if a law is universal why shouldn't it apply to the universe? It remains a great temptation to take a law or principle we can successfully apply to all the world's subsystems and apply it to the universe as a whole. To do so is to commit a fallacy I will call the *cosmological fallacy*.

The universe is an entity different in kind from any of its parts. Nor is it simply the sum of its parts. In physics, all properties of objects in the universe are understood in terms of relationships or interactions with other objects. But the universe is the sum of all those relations and, as such, cannot have properties defined by relations to another, similar entity.

Thus the Earth is, in Anaximander's universe, the one thing that doesn't fall, because it is the thing that objects fall to. Similarly, our universe is the one thing that cannot be caused by or explained by something external to it, because it is the sum of all the causes.

If the analogy of the present period to ancient Greek science is apt, there will be paradoxes and unanswerable questions that follow from the act of extending small-scale laws to the universe as a whole. There are both. We, in our time, are led by our faith in the Newtonian paradigm to two simple questions that no theory based on the that paradigm will ever be able to answer:

- *Why these laws?* Why is the universe governed by a particular set of laws? What selected the actual laws from other laws that might have governed the world?
- The universe starts off at the Big Bang with a particular set of initial conditions. *Why these initial conditions?* Once we fix the laws, there are still an infinite number of initial conditions the

universe might have begun with. What mechanism selected the actual initial conditions out of the infinite set of possibilities?

The Newtonian paradigm cannot even begin to answer these two enormous questions, because the laws and initial conditions are inputs to it. If physics ultimately is formulated within the Newtonian paradigm, these big questions will remain mysteries forever.

We used to think we knew how the *Why these laws?* question would be answered. Many theorists believed that only one mathematically consistent theory could incorporate the four fundamental forces of nature — electromagnetism, the strong and weak nuclear forces, and gravity, within quantum theory. Had this been the case, the answer to the *Why these laws?* question would have been that only one possible law of physics could give rise to a world roughly like ours.

But this hope was dashed. By now, we have good evidence that there is no unique theory incorporating everything we know about nature — in essence, a theory reconciling general relativity and quantum mechanics. There has been a great deal of progress in the last thirty years on several different approaches to quantum gravity, and they lead to the conclusion that to the extent each succeeds, it does so in a way that is not at all unique. The best-studied approach to quantum gravity is loop quantum gravity, and it appears to allow a wide range of choices of elementary particles and forces.

The same is true of string theory, which also promises a unification of gravity and quantum theory. There is evidence for the existence of an infinite number of string theories, many of which depend on a large set of parameters — numbers that can be tuned by hand to any values we choose. All these theories appear equally consistent mathematically. A vast number of them describe worlds with a spectrum of elementary particles and forces roughly like ours — although, as of now, no string theory has been constructed that precisely includes the Standard Model of Particle Physics.

The original hope of string theory had been that there would be a unique fundamental theory that would precisely reproduce the Standard Model and give specific predictions for observations beyond it.

In 1986, Andrew Strominger discovered that string theory comes in a vast number of versions, killing this hope.[2] This is what impelled me to wonder how the universe might have selected its laws — and led to my eventual embrace of the reality of time.

So much for unanswerable questions. What about dilemmas?[3] As it happens, a big one lies at the heart of the usual notion of a law of physics as expressed by the Newtonian paradigm. What we mean when we call something a "law" is that it applies to many cases; if it applied to just one, it would simply be an observation. But any application of a law to any part of the universe involves an approximation, as we saw in chapter 4, because we must neglect all the interactions between that part and the rest of the universe. So the many applications of a law of nature that are checkable are all approximations.

To apply a law of nature without approximation, we must apply it to the whole universe. But there is only one universe — and one case does not yield sufficient evidence to justify the claim that a particular law of nature applies. This might be called the cosmological dilemma.

The cosmological dilemma need not prevent us from applying laws of nature — like general relativity, or Newton's laws of motion — to subsystems of the universe. They work in virtually all cases, and this is why we call them laws. But each such application is an approximation based on the fiction of treating a subsystem of the universe as if it were all there is.[4] Nor does it prevent us from imagining that the history of our universe is a solution to a law like general relativity, with matter described by the Standard Model. But it doesn't explain why that solution was picked as the only one to be realized in nature. Nor does one solution prove that the laws of nature that do exist are a combination of general relativity and the Standard Model, because that one solution could approximate solutions to many different laws.[5]

Here's an example to illustrate why a law has to be testable in more than one case to be distinguishable from simple observation. A family has one child, Mira, who loves ice cream. Her favorite is chocolate — indeed, the first ice cream she ever tasted was chocolate, and she has preferred it ever since.

Mira's parents believe there is a general law that all children love ice cream. But without observing any other children, they have no way to test this, no way to distinguish this from what they have observed, which is that Mira loves ice cream. Mira's father also believes in another law, which is that all children prefer chocolate ice cream. As they relax with a drink after Mira is in bed, her mother counters with yet another hypothesis: All children prefer the kind of ice cream they first taste.

Both possibilities are consistent with the evidence they have. They make different predictions, which could be tested by polling parents in the playground, hence both are possible laws. But suppose that Mira is the only child that exists. There will be no way to test whether any of her parents' hypotheses are general laws or just observations.

Mira's parents might argue, based on human biology, that children will love anything made from sugar and milk and this validates at least one of their predictions. They would be correct, but their reasoning uses knowledge gained from studies of many humans. This is where the analogy breaks down, because in cosmology there is genuinely only one case. In a scientific argument, the universe cannot be presumed to be a single case of a more general class, because no assertion as to the characteristics of that class are testable.

The point that laws applying to subsystems must be approximate is central to the cosmological dilemma, so let me abandon ice cream and give a simple example of this point from physics. Newton's first law of motion asserts that all free particles move along straight lines. It has been tested and confirmed in numerous cases. But each test involves an approximation, because no particle is truly free. Every particle in our universe feels a gravitational force from every other. If we wanted to check the law exactly, there would be no cases to apply it to.

Newton's first law can, at best, be an approximation to some more exact other law. Indeed, it approximates Newton's second law, which describes how a particle's motion is influenced by the forces it experiences. Now, here is something very interesting! Each particle in the universe attracts every other gravitationally. There are also forces between every pair of charged particles. That's a whole lot of forces to contend with. To check whether Newton's second law holds exactly, you would have to add up more than 10^{80} forces to predict the motion of only one of the particles in the universe.

In practice, of course, we do nothing of the kind. We take into account just one or a few forces from nearby bodies and ignore all the rest. In the case of gravity, for example, we can justify neglecting forces from far-distant bodies because their influences are weaker. (This is not as obvious as it sounds, because although the forces from far-away particles are weaker, there are many more faraway particles than nearby ones.) In any event, no one ever attempts to check whether the second law is *exactly* true. We check only extreme approximations to it.

Another big problem with extrapolating the Newtonian notion of "law" to the whole universe is that there is one universe but an infinite choice of initial conditions. These correspond to an infinite number of solutions to the equations of an alleged cosmological law — solutions that describe an infinite set of possible universes. But there is only one actual universe.

The very fact that a law has an infinite number of possible solutions describing an infinite number of possible histories forces us to conclude that it is meant to be applied to subsystems of the universe, which come, in nature, in an enormous number of versions. The plentitude of nature is matched by the plentitude of solutions. So when we apply a law to a small subsystem of the universe, the freedom to specify initial conditions is a necessary part of the law's success.

By the same token, when we apply a law that has an infinite number of solutions to a unique system, such as the universe, we leave a great deal unexplained. The freedom to choose the initial conditions turns from an asset into a liability, because it means that there are essential questions about the one universe that the theory (which that law expresses) gives no answer to. These include any feature of the universe that depends on the universe's initial conditions.

What are we to think about all the other histories that are also solutions of the alleged cosmological law, but that the universe does not follow? Why the vast extravagance of an infinite number of solutions, only one of which, at most, could have anything to do with nature?

These considerations point to one conclusion: We are mistaken about what a law of nature could be on a cosmological scale. For three reasons:

(1) The assertion that a law applies on a cosmological scale implies a vast amount of information about predictions concerning nonexistent cases — that is, other universes. This suggests that something much weaker than a law might explain the universe. We don't need an explanation so extravagant that it makes predictions about an infinite number of cases that never happen. An explanation that accounts only for what actually happens in our single universe would suffice.

(2) The usual kind of law cannot explain why the solution that describes our universe is the one we experience.

(3) The law cannot account for itself. It offers no rationale for why it, rather than some other law, holds.

So a conventional natural law, applied to the universe, explains at the same time way too much and not nearly enough.

The only way to escape these dilemmas and paradoxes is to seek a methodology that goes beyond the Newtonian paradigm — a new paradigm, applicable to physics on the scale of the universe. Unless we are content to let physics end in irrationality and mysticism, the method that is the basis of its success so far must be superseded.

But all the arguments set out in Part I for the expulsion of time from physics were based on the assumption that the Newtonian paradigm can be extended to the universe as a whole. If it can't, then those arguments for the elimination of time fail. When we abandon the Newtonian paradigm, we must abandon those arguments, and it becomes possible to believe that time is real.

Can we do better at making a true cosmological theory if we embrace the reality of time? In the following chapters, I will explain why the answer is yes.

9

The Cosmological Challenge

THE GREAT THEORIES of 20th-century physics — relativity, quantum theory, and the Standard Model — represent the highest achievements of physical science. They have beautiful mathematical expressions that result in precise predictions for experiments, which have been confirmed in many cases to great accuracy. And yet I have just argued that nothing along the lines of these theories can serve as a fundamental theory. This is an audacious claim in the light of their success.

, To support this claim, I can point to a feature that all established theories of physics share and which makes it difficult to extend them to the whole universe: Each divides the world into two parts, one that changes over time and a second assumed to be fixed and unchanging. The first part is the system being studied, whose degrees of freedom change with time. The second part is the rest of the universe; we can call it the background.

That second part may not be described explicitly, but it is implicitly there in the terms that give meaning to the motion described in the first part. A distance measurement implicitly refers to the fixed points and rulers needed to measure that distance; a specified time implies the existence of a clock outside the system measuring the time.

As we saw in the game of catch in chapter 3, the position of the ball is made meaningful by reference to where Danny is standing. The motion is defined using a clock, which is assumed to tick at a constant rate. Both Danny and the clock are outside the system described by the configuration space and are assumed to be static. Without these fixed reference points, we would not know how to connect predictions of the theory with records of experiments.

This division of the world into a dynamical and a static part is a fiction, but it is an extremely useful one when it comes to describing small parts of the universe. The second part, assumed to be static, in reality consists of other dynamical entities outside the system being analyzed. By ignoring their dynamics and evolution, we create a framework within which we discover simple laws.

For most theories except general relativity, the fixed background includes the geometry of space and time. It also includes the choice of laws, as these are assumed to be changeless. Even general relativity, which describes a dynamical geometry, assumes other fixed structures, such as the topology and dimension of space.[1]

This division of the world — into a dynamical part and a background that surrounds it and defines the terms with which we describe it — contributes to the genius of the Newtonian paradigm. But it is also what renders the paradigm unfit for application to the whole universe.

The challenge we face when extending science to a theory of the whole universe is that there can be no static part, because everything in the universe changes, and there is nothing outside of it — nothing that can serve as a background against which to measure the motion of the rest. The invention of a way to surmount this barrier might be called the *cosmological challenge.*

The cosmological challenge requires us to formulate a theory that can be applied meaningfully to the whole universe. It must be a theory in which every dynamical actor is defined in terms only of other dynamical actors. Such a theory has no need of, and no place for, a fixed background; we call such theories background-independent.[2]

We can see now that the cosmological dilemma is built into the structure of the Newtonian paradigm, because the very features responsible for success on smaller scales — including the dependence on fixed backgrounds and the fact that one law has an infinite number of solutions — turns into the reason for the paradigm's failure as the basis for a theory of cosmology.

We are fortunate to live in a time when the success of physics has led to the first attempts to study cosmology scientifically. It's not surprising that one response to the cosmological dilemma is to posit that the universe is one of a vast collection, because all our theories can be applied only to a part of a vastly larger system. This, at any rate, is how I understand the attraction of the various multiverse scenarios.

❖

When we do an experiment in a laboratory, we control the initial conditions. We vary them to test hypotheses about laws. But when it comes to cosmological observations, the initial conditions were set in the early universe, so we must make hypotheses as to what those were. Thus, to explain the result of a cosmological observation using the Newtonian paradigm, we make *two* hypotheses: We hypothesize what the initial conditions were *and* what laws operated on them. This puts us in a much more challenging situation than the usual context of physics in a box, in which we use the control we have over the initial conditions to test hypotheses as to the laws.

The fact that we must simultaneously test hypotheses about the laws and the initial conditions greatly weakens how well we can test either. If we make a prediction and observation disagrees, there are two ways to correct the theory: We can modify our hypothesis about the laws, or we can modify our hypothesis about the initial conditions. Each can affect the results of the experiment.

This raises a new problem, for how do we know which of the two hypotheses needs to be corrected? If the observation is of a small part of the universe, like a star or a galaxy, we base our testing of the law on an analysis of many cases. They were all subject to the same law, so any differences between them have to be attributed to differences in their initial conditions. But where the universe is concerned, we cannot distinguish the effects of changing the hypothesis about a law from the effects of changing the hypothesis about the initial conditions.

This problem occasionally intrudes on cosmological research. A major test of theories of the early universe is to account for the patterns seen in the cosmic microwave background (CMB). This is radiation left over from the early universe which gives us a snapshot of conditions about 400,000 years after the Big Bang. One much studied proposal is cosmological inflation, which posits that very early in its history the universe underwent a tremendous and rapid expansion. This stretched out and reduced whatever its initial features were and led to the big, relatively featureless universe we observe. Inflation also predicts patterns in the CMB very similar to what has been seen.

A few years ago, observers reported evidence for a new feature in the microwave background, non–Gaussianity, that is not predicted by the usual theory of inflation.[3] (It doesn't matter what non–Gaussianity is; all we need to know for this story is that it's a pattern that may have been observed in the CMB and that the standard theory of inflation predicts should not occur.) We have two options to explain the new observation: We can modify the theory or we can modify the initial conditions.

The theory of inflation falls into the Newtonian paradigm, so its predictions depend on the initial conditions that the laws act on. Within days of the first paper presenting evidence for non–Gaussianity, papers appeared attempting to explain the observation. Some modified the laws, others modified the initial conditions. Both strategies succeeded in retrodicting the claimed observations — indeed, that either strategy would work was already known.[4] As is typical on the frontier of observational science, further observations have failed to support the initial claim. As of this writing, we still don't know if there really is non–Gaussianity in the CMB.[5]

This is an example of a case in which there are two different ways to fit a theory to the data. If we consider that the laws and initial conditions are described by some parameters, there are two distinct parameter fits to the observed data. Observers call this kind of situation a *degeneracy*. Usually when there is a degeneracy, we make additional observations to distinguish which fit is correct. But in a case like the CMB, which is the remnant of an event that happened only once, we may never be able to resolve the degeneracy. Given especially the limits to how well we can measure the CMB, it's possible that we may not be able to disentangle an explanation based on modifying the laws from one based on modifying the initial conditions.[6] But without the ability to disentangle the role of laws and initial conditions, the Newtonian paradigm loses its power to explain the causes of physical phenomena.

◆

We are ready to reverse the expectations that have guided physics from the time of Newton until very recently. Formerly, we thought of theories like Newtonian mechanics or quantum mechanics as candidates for fundamental theories that — if they succeeded — would be perfect mirrors of the natural world, in the sense that everything true about nature would be echoed by a mathematical fact that was true of the theory. The very structure of the Newtonian paradigm, based on timeless laws acting on a timeless space of configurations, was thought to be essential to this mirroring. I am proposing that this aspiration was a metaphysical fantasy guaranteed to lead to the aforementioned dilemmas and confusions as soon as we tried applying that paradigm to the whole universe. This stance requires a re-evaluation of the status of theories within the Newtonian paradigm — from candidates for fundamental theories to approximate descriptions of small subsystems of the universe. It is a re-evaluation that has already taken place among physicists and consists of two related changes of perspective:

(1) All the theories we work with, including the Standard Model of Particle Physics and general relativity, are approximate theories, applying to truncations of nature that include only a subset of the degrees of freedom in the universe. We call such an approximate theory an *effective theory.*

(2) In all our experiments and observations involving truncations of nature, we record the values of a subset of degrees of freedom and ignore the rest. The resulting records are compared with the predictions of effective theories.

So the success of physics to this day is entirely based on the study of truncations of nature, which are modeled by effective theories. The art of doing physics at the experimental level is all about designing experiments to isolate and study a few degrees of freedom, ignoring the rest of the universe. The methodologies of theorists are aimed at inventing effective theories to model the truncations of nature that the experimentalists study. Never in the history of physics have we been able to compare the predictions of a candidate for a truly fundamental theory — by which I mean one that cannot be understood as an effective theory — with experiment.

Let me explain a bit more about these points:

Experimental physics is the study of truncations of nature.

A subsystem of the universe modeled as if it were the only thing in the universe, neglecting everything outside it, is called an *isolated system.* But we should never forget that isolation from the whole is never complete. As noted, in the real world there are always interactions between any subsystem we may define and things outside it. To one extent or another, subsystems of the universe are always what physicists call *open systems.* These are bounded systems that interact with things beyond those boundaries. So when we do physics in a box, we are approximating an open system by an isolated system.

A great deal of the craft of experimental physics consists of turning open systems into (approximately) isolated systems. We can never do this perfectly. For one thing, the measurements we make on the system intrude upon it. (This is a big issue in the interpretation of quantum mechanics; but for now let's stick to the macro world.) Every experiment is a fight to extract the data you want from the unavoidable presence of noise coming from outside your imperfectly isolated system. Experimenters spend a lot of effort convincing themselves and their peers that they are seeing a real signal standing out from the noise, and we do what we can to reduce its effect.

We shield our experiments from contamination from outside vibrations, fields, and radiation. For many experiments, this suffices, but some experiments are so delicate that they are affected by noise from cosmic rays hitting the detectors. To do a good job of isolating a laboratory from cosmic rays, you can put it in a mine several miles underground; this is what we do to detect neutrinos from the sun. It reduces the random background noise from other radiation to manageable numbers, while the neutrinos still get through. But there's no practical way to isolate a laboratory from neutrinos. Neutrino detectors buried deep in the ice at the South Pole record neutrinos that have entered at the North Pole and traveled all the way through the Earth.

Even if you could build walls astronomically thick of dense lead to screen out neutrinos, there's something that still would get through, and that is gravity. In principle, nothing can screen out the force of gravity or stop the propagation of gravitational waves, so nothing can be perfectly isolated. I discovered this important point during my PhD studies. I wanted to model a box that contained gravitational waves bouncing back and forth inside, but my models kept failing, because the gravitational waves passed right through the walls. I imagined increasing the density of the walls of a box higher and higher, to the point where they would reflect gravitational radiation, but before I reached that point the model showed the material in the walls collapsing into black holes. After banging my head against this for some time, trying one way and then another to avoid it, I eventually realized that the obstacle I couldn't overcome was itself a much more interesting discovery than the one I was trying to make work. With some further thought, I was able to show that given only a few assumptions, no wall can screen out gravitational waves.[7] No matter what the wall is made of or how thick or dense it is. To reach this conclusion, I had to assume only the laws of general relativity, that the energy contained in matter is positive, and that sound cannot travel faster than light.

This means — not just in practice but also in principle — that there's no such thing as a system in nature that is isolated from influence by the rest of the universe. It's worth elevating this conclusion to a principle, which I will call the *principle of no isolated systems.*

There's another reason a model of an open system as an isolated system is always an approximation, which is that we can't anticipate random disruptive incursions. We can measure, anticipate, and deal with the noise. But the outside world can do far worse to our attempts to isolate our system. A plane may crash into the building housing our laboratory, or an earthquake may topple it. An asteroid may collide with Earth. A cloud of dark matter may pass through the solar system, perturbing Earth's orbit and plunging us into the sun.[8] Or maybe someone will turn off the power in the lab by pulling a switch in the basement. The list of things that might happen in this big universe of ours to disrupt an experiment in progress is virtually endless. When we model an experiment as if it were an isolated system, we are excluding all these possibilities from our model.

To incorporate everything that might impinge on our laboratory from the outside would require a model of the whole universe. We can't do physics without excluding all these possibilities from our models and calculations. Yet to exclude them is, in principle, to base our physics on approximations.

Effective but approximate theories.

All the major theories of physics are models of the truncations of nature produced by experimenters. They may have been imagined as fundamental theories when they were invented, but over time theorists have come to understand them as effective descriptions of a limited number of degrees of freedom.

Particle physics provides a good example of the role of effective theories. Experiments up to now have probed fundamental physics only down to a certain length scale. Currently this is about 10^{-17} of a centimeter, which is probed by the Large Hadron Collider at CERN. This means that the Standard Model of Particle Physics, which agrees with all known experiments so far, must be considered an approximation (besides the fact that it has nothing to say about gravity). It ignores currently unknown phenomena that might appear were we able to probe to shorter distances.

There is an inverse relationship between length scale and energy in quantum physics, due to the uncertainty principle. To probe down to a certain length scale, we need particles or radiation of at least a certain energy. To go to shorter distances, we need higher-energy particles. So the lower limit on length scales we have reached is based on an upper limit on the energies of processes we have observed. But since energy and mass are the same thing (according to special relativity), this means that if we have probed only up to a certain energy scale, we could be ignorant of particles too massive to have been created by our collider experiments thus far. The missing phenomena could include not only new kinds of elementary particles but also heretofore unknown forces. Or it could turn out that the basic principles of quantum mechanics are wrong and need modification to correctly describe phenomena lurking at shorter lengths and higher energies.

Because of these concerns, we speak of the Standard Model as an effective theory, one compatible with experiment but reliable only within a certain domain.

The notion of effective theories subverts some well-worn notions, such as the platitude that simplicity and beauty are hallmarks of truth. Since we don't know what could be lurking at higher energies, many hypotheses of physics beyond its specified domain are consistent with one or another effective theory. So these effective theories have an intrinsic simplicity, because they have to be consistent with the simplest and most elegant way they could be extended into unknown domains. A large part of the elegance of general relativity and the Standard Model is explained by understanding them as effective theories. Their beauty is a consequence of their being effective and approximate. Simplicity and beauty, then, are the signs not of truth but of a well-constructed approximate model of a limited domain of phenomena[9].

The notion of an effective theory represents a maturing of the profession of elementary-particle theory. Our young, romantic selves dreamed we had the fundamental laws of nature in our hands. After working with the Standard Model for several decades, we are now simultaneously more confident that it's correct within the limited domain in which it has been tested and less confident of its extendability outside that domain. Isn't this a lot like real life? As we grow older, we gain confidence about what we really know and simultaneously find it easier to admit ignorance about what we don't.

This may seem disappointing. Physics is supposed to be about discovering the fundamental laws of nature. An effective theory is by definition not that. If you have too naïve a view of science, you might think that a theory could not both agree with all experiments yet carried out and be considered at best only an approximation to the truth. The concept of an effective theory is important, because it expresses this subtle distinction.

It also exemplifies how we understand progress in elementary-particle physics. It tells us that physics is a process of constructing better and better approximate theories. As we push our experiments to shorter distances and larger energies, we may discover new phenomena, and if we do, we'll need a new model to accommodate them. Just like the Standard Model, it will be an effective theory, albeit one applicable in a wider domain.

The notion of an effective theory implies that progress in physics entails revolutions that completely change the conceptual basis of our understanding of nature while preserving the successes of earlier theories. Newtonian physics can be considered an effective theory, applying to a domain in which speeds are much lower than that of light and quantum effects can be ignored. Within that domain, it remains as successful as it ever was.

General relativity is another example of a theory that was once a candidate for a fundamental description of nature but which is now understood to be an effective theory. For one thing, it leaves out the domain of quantum phenomena. General relativity is, at best, an approximation to a unified quantum theory of nature, and may be arrived at by truncating that more fundamental theory.

Quantum mechanics, too, is likely an approximation to a more fundamental theory. One sign of this is the fact that its equations are linear — meaning that the effects are always directly proportional to their causes. In every other example in which a linear equation is used in physics, the theory is known to arise as an approximation to a more fundamental (but still effective) theory that is nonlinear (in the sense that the effects may be proportional to a higher power of the cause), and the best bet is that this will turn out to be true of quantum mechanics as well.

The fact is that every theory we have so far used in physics has been an effective theory. It is sobering to realize that part of the cost of their success was the realization that they are approximations.

We still may harbor the ambition to invent a fundamental theory that describes nature without approximation. Both logic and history tell us this is impossible as long as we stay within the Newtonian paradigm. So as admirable as Newtonian physics, general relativity, quantum mechanics, and the Standard Model are, they cannot be the template for a fundamental cosmological theory. The only possible path to such a theory is to take up the cosmological challenge and devise a theory not patterned on the Newtonian paradigm that can be applied to the whole universe without approximation.

10

Principles for a New Cosmology

E NOW BEGIN OUR SEARCH for a theory that can truly be a theory of the whole universe. Such a theory must avoid the cosmological dilemma, and it must also be background-independent — not presuming a division of the world into two parts, one containing dynamical variables that evolve, the other containing fixed structures providing a background to give meaning to the evolving parts. Everything that the theory asserts is part of reality must be defined by its relationships to the rest of reality, in a way that renders it subject to change.

What must we require of a true cosmological theory?

- *Any new theory must contain what we already know about nature.* We need the current theories — the Standard Model of Particle Physics, general relativity, and quantum mechanics — to emerge as approximations to the unknown cosmological theory whenever we restrict our attention to scales of distance and time smaller than the cosmos.
- *The new theory must be scientific.* Genuine explanations show

their validity by having myriads of unanticipated consequences. There can be no just making things up because it makes a nice story. A real theory must imply specific testable predictions.

- *The new theory should answer the "Why these laws?" question.* It must give us substantial insight into how and why the particular elementary particles and forces described in the Standard Model were selected. In particular, it must explain the special and improbable values of the fundamental constants that obtain in our universe—the parameters, like the masses of the elementary particles and the strengths of the various forces, that are specified by the Standard Model.
- *The new theory should answer the "Why these initial conditions?" question,* explaining why our universe has properties that seem unusual when compared to the possible universes that might be described by the same laws.

These are minimal requirements. Given that we are speaking of a theory of the whole universe, the collective wisdom of physics — contained in the writings of the great minds that have struggled to invent theories of the natural world, among them Kepler, Galileo, Newton, Leibniz, Ernst Mach, and Einstein — dictates that we can specify a few more.[1] Here's my interpretation of what these sages have taught us:

The explanations such a theory gives of features of our universe must depend only on things that exist or occur within the universe. No chains of explanation can point outside the universe. So we must demand a *principle of explanatory closure.*

To be scientific, a theory needn't give the precise answer to any question you can conceive of, but there should be a great number of questions we believe we could answer if we knew more details about the universe. Leibniz's *principle of sufficient reason* postulates that there should be an answer to any reasonable question we might ask about why the universe has some particular feature. An important test of a new scientific theory is whether it increases the number of questions we can answer. Progress occurs when we discover reasons for aspects of the universe that were unexplained by earlier theories.

Leibniz's principle has some consequences that should constrain a cosmological theory. One is that there should be nothing in the universe that acts on other things without itself being acted upon. All influences or forces should be mutual. We can call this the *principle of no unreciprocated actions*. Einstein invoked this principle to justify his replacement of Newton's theory of gravity by general relativity. His point was that Newton's absolute space tells bodies how to move, but nothing is reciprocated; the bodies in the universe do not influence absolute space. Absolute space just is. In Einstein's theory of general relativity, the relationship between matter and geometry is reciprocal: Geometry tells matter how to move and, in turn, matter influences the curvature of spacetime. Nor can anything affect the flow of Newton's absolute time. Newton hypothesizes that it flows the same whether the universe is empty or full of matter. In general relativity, the presence of matter affects the behavior of clocks.

This principle, then, forbids any reference to fixed-background structures — entities whose properties are fixed for all time, regardless of the motion of matter.

These background structures are the unconscious of physics, silently shaping our thinking to give meaning to the basic concepts we use to imagine the world. We think we know what "position" means because we are making unconscious assumptions about the existence of an absolute reference. Several of the fundamental steps in the evolution of physics have consisted of recognizing the existence of a fixed-background structure, removing it, and replacing it with a dynamical cause within the universe. This was what Ernst Mach did when he refuted Newton by suggesting that we feel dizzy when we spin because we move relative to the matter in the universe rather than to absolute space.

If we insist on reciprocal action and rule out fixed-background structures, what we are saying is that every entity in the universe evolves dynamically, in interaction with everything else. This is the essence of the philosophy of relationalism, which is usually attributed to Leibniz (recall our discussion of the meaning of "position" in chapter 3). We can extend this idea to assert that all properties in a cosmological theory should reflect evolving relationships among dynamical entities.

But if the properties of a body — the properties by which we identify it and distinguish it from other bodies — are relationships with other bodies, then there cannot be two bodies that have the same set of relationships to the rest of the universe. Two things that have the same relationships with everything else in the universe must actually be the same thing. This is another of Leibniz's principles, called the *identity of the indiscernibles.* It too is a consequence of the principle of sufficient reason, for if there are two distinct entities with the same relationships to the rest of the world, there is no reason they should be as they are and not exchanged. This would amount to a fact about the world that had no rational explanation.

So there can be no fundamental symmetries in nature. A symmetry is a transformation of a physical system that exchanges its parts while leaving all its physically observable quantities the same.[2] An example of a symmetry of Newtonian physics is translation of a subsystem from one place in space to another. Since the laws of physics don't depend on where a system is in space, the predictions are unchanged if the laboratory — and everything that may affect the experimental results — is moved six feet to the left. We state the independence of experimental results from location in space by saying that physics is invariant under translating systems in space.

Symmetries are common in all the physical theories we know. Several of the most useful tools in the physicist's toolbox exploit the presence of symmetries. Yet if Leibniz's principles are right, they must not be fundamental.

Symmetries arise from the act of treating a subsystem of the universe as if it were the only thing that existed. It is only because we ignore the interactions between the rest of the universe and the atoms in our laboratory that it doesn't matter if we move the laboratory in space. This also explains why it doesn't matter if we rotate the subsystem we're studying. It doesn't matter because we ignore the interac-

tions between that subsystem and the rest of the universe. If we took
those interactions into account, it certainly would matter if the subsystem were rotated.

But what if the universe itself is translated or rotated? Isn't that a
symmetry? It is not, because no relative position within the universe is
altered. In the relational perspective, there is no meaning to translating or rotating the universe. Symmetries, such as translations and rotations, are then not fundamental; they arise from the division of the
world into two parts, as described in the preceding chapter. These and
other symmetries are features only of approximate laws applying to
subsystems of the universe.

This has a stunning consequence: If these symmetries are approximate, then so are the laws of conservation of energy, momentum, and
angular momentum. These basic conservation laws depend on the
assumption that space and time are symmetric under translations in
time, translations in space, and rotations. The connection between
symmetries and conservation laws is the content of a basic theorem
proved early in the 20th century by the mathematician Emmy Noether.[3] I won't try explaining her reasoning here, but her theorem is
one of the pillars of physics and deserves to be better known.

So the unknown cosmological theory will have neither symmetries
nor conservation laws.[4] Some particle physicists, impressed by the
success of the Standard Model, like to say that the more fundamental
a theory is, the more symmetries it should have. This is precisely the
wrong lesson to draw.[5]

◆

We now come to the most important question about the unknown
cosmological theory: What will it have to say about the nature of time?
Will time be dissolved, as in Einstein's theory of general relativity? Will
time disappear and emerge only when needed, as in Barbour's quantum cosmology? Or will time play an essential role, unlike any of the
theories since Newton?

I believe that time is needed for any theory that answers the *Why these laws?* question. If laws are to be explained, they must evolve. This was argued by Charles Sanders Peirce, whom I quoted in the Introduction. Let's look at that quote again, to untangle the argument he's making. He begins, "To suppose universal laws of nature capable of being apprehended by the mind and yet having no reason for their special forms, but standing inexplicable and irrational, is hardly a justifiable position."

We can understand this as a statement of Leibniz's principle of sufficient reason: We should be able to say why the laws of nature we have discovered, rather than others, are the laws. Peirce re-emphasizes this in the following two sentences: "Uniformities are precisely the sort of facts that need to be accounted for. . . . Law is par excellence the thing that wants a reason."

This is a statement of the *Why these laws?* question. Facts about the world need to be explained, and a fact most in need of explanation is why particular laws are observed to hold in our universe.

Then he asserts that "the only possible way of accounting for the laws of nature and for uniformity in general is to suppose them results of evolution." This is a strong statement. Peirce makes no argument for his conclusion that laws must evolve; he simply claims that it is the "only possible" solution to the *Why these laws?* question. I don't know whether, anywhere in his voluminous writings and notebooks, he ever provided an argument for that conclusion. But here's one he might have made.

Our task is to explain why an object — in this case, the universe — has a particular property: namely, that the elementary particles and forces interact through processes described by the Standard Model of Particle Physics. The problem is challenging, because we know that the Standard Model, with its particular parameters, is just one of a huge number of possible choices for the laws of nature. So how do we explain why an entity has a particular property out of a vast set of possible alternatives?

Since there are many alternatives, no principle specifies the precise laws we see. If there's no necessary reason for the choice, then there must be some reason that falls short of logical necessity. There could be, or could have been, cases in which the choice was made differently. How do we explain how the choice was made in our universe's case?

If there really is just the one case, there will never be a sufficient explanation, because, ipso facto, there's no logical principle that determines the choice. A sufficient explanation requires there to have been other universes initially endowed with laws. That is, there must have been more than one event like our Big Bang in which laws of nature were chosen. (For simplicity's sake, we're assuming that laws are chosen at dramatic events like our Big Bang; we certainly have no evidence that the laws of nature have changed since then.)

The question then is how the Big Bangs — the law-choosing events — are arranged. We can now invoke the principle that the universe must be explanatorily and causally closed. That is, we assume that the universe contains all the chains of causes necessary to explain anything within it. If we want to explain how the effective laws were chosen at our Big Bang, we can invoke only events in the past of the Big Bang. And we can apply the same logic to the causes of the choices of laws made at Bangs prior to ours. There must thus be a sequence of Bangs extending endlessly back into the past. Let's pick an arbitrary starting point many Bangs ago and follow the choices of laws forward. We will see the laws evolving as our present universe is approached. So we reach Peirce's conclusion that if we hope to explain laws, those laws must have evolved.[6]

The Bangs may be purely sequential or they may branch — into the future or the past or both. We can construct different hypotheses as to whether there's a branching and as to what exactly takes place at these events to modify the laws of nature. In all these cases, we will be explaining the choice of laws made in the most recent Big Bang in terms only of events in its causal past. A scenario of this kind might well be checked experimentally; the events before our Big Bang might be observable through information left in remnants (if any) that survived the birth of our universe. In chapters 11 and 18, we'll see examples of predictions made by theories that allow the laws of nature to evolve before our Big Bang.

However, if the Big Bang has no past, the choice of laws and initial conditions is arbitrary and there will be no such tests. Nor will there be any tests of scenarios in which a vast or infinite population of universes exist whose Big Bangs are all causally disconnected from ours. In a scientific cosmology, the postulation of *parallel universes,* universes that are causally unconnected to ours, cannot help us explain any property of our own universe. We conclude that the only way to have a scientific cosmological theory that can make falsifiable predictions is if the laws evolved in time. (A prediction of a theory is falsifiable if it could be contradicted by a doable experiment.)

Roberto Mangabeira Unger puts this more elegantly.[7] Either time is real or it is not. If time is not real, then laws are timeless — but then the choice of laws is inexplicable, for reasons we have already discussed. If, on the other hand, time is truly real, then nothing, not even the laws, can last forever. If the laws of nature act forever, we are in the Newtonian paradigm, and you could use them to reduce any property of the world at a later time to a property at an earlier time. Or, equivalently, you could replace any physical causation with logical implication. So time being real means laws don't last forever. They must evolve.

The notion of timeless laws also violates the relational principle that nothing in the universe acts without being acted on. If you choose to except the laws of nature from this principle, seeing them as something outside the universe, you put them outside the realm of rational explanation. To make laws explicable, we must consider them as much a part of the world as the particles they act on. This brings them into the purview of change and causality. They become explicable only when they participate in the dance of change and mutual influence that makes the world a whole.

Although we don't yet have the cosmological theory, we already know something about it, if the principles I've put forward are sound:

- It should contain what we already know about nature, but as approximations.
- It should be scientific; that is, it has to make testable predictions for doable experiments.
- It should solve the *Why these laws?* problem.

- It should solve the initial-conditions problem.
- It will posit neither symmetries nor conservation laws.
- It should be causally and explanatorily closed. Nothing outside the universe should be required to explain anything inside the universe.
- It should satisfy the principle of sufficient reason, the principle of no unreciprocated action, and the principle of the identity of the indiscernibles.
- Its physical variables should describe evolving relationships between dynamical entities. There should be no fixed-background structures, including fixed laws of nature. Hence the laws of nature evolve, which implies that time is real.

Principles are well and good, but what we really need are hypotheses that lead to theories that make testable predictions. In the next several chapters, I will describe several examples of hypotheses and theories that realize these principles, and we will see that they do indeed make testable hypotheses.

11

The Evolution of Laws

T HE MAIN MESSAGE OF Part II so far has been that for cosmology to progress, physics must abandon the idea that laws are timeless and eternal and embrace instead the idea that they evolve in a real time. This transition is necessary so that we can arrive at a cosmological theory — one that explains the choices of laws and initial conditions — that is testable and even vulnerable to falsification by doable experiments. Having (I hope) made the case for this in principle, I'll demonstrate it in this chapter, by comparing the ability of two theories, one timeless and one with evolving laws, to explain and predict observational results.

The theory in which laws evolve is called *cosmological natural selection,* which I developed in the late 1980s and published in 1992.[1] In that paper, I made a few predictions, which could have been falsified in the two decades since but have not been. This of course doesn't prove the theory is correct, but at least I showed that a theory of evolving laws can explain and predict real features of our world.

For an example of a timeless theory, I will take a version of the multiverse scenario called *eternal inflation,* proposed in the 1980s by Al-

exander Vilenkin and Andrei Linde and widely studied since.[2] Eternal
inflation comes in different forms, reflecting the fact that some of its
hypotheses are adjustable. To make my point, I've chosen one simple
form that best fits the "eternal" because it gives a timeless picture of
the multiverse. There are other versions of inflationary multiverses in
which time plays a more essential role, and to the extent that these
involve a genuine notion of evolving laws they share some aspects of
cosmological natural selection.

One reason that cosmological scenarios with evolving laws succeed
in making real predictions is that they don't rely on the *anthropic principle* — which states that we can live only in a universe whose laws and
initial conditions create a universe hospitable to life — to connect the
multiverse with the universe we observe. One of the tasks of this chapter is to refute the claim that the anthropic principle can play a role in
making a theory predictive.

Cosmological natural selection was the subject of my first book,
The Life of the Cosmos, so I will describe it in just enough detail to
make clear why evolution of the laws in time leads to a falsifiable explanation of them.[3]

The basic hypothesis of cosmological natural selection is that universes reproduce by the creation of new universes inside black holes.
Our universe is thus a descendant of another universe, born in one of
its black holes, and every black hole in our universe is the seed of a
new universe. This is a scenario within which we can apply the principles of natural selection.

The mechanism of natural selection I use is based on the methods of population biology that serve to explain how some parameters
governing a system can be selected that make it more complex than it
would otherwise be. Applying natural selection to a system to explain
its complexity requires the following:

- *A space for parameters that vary among a population.* In biology, these parameters are the genes. In physics, they are the constants of the Standard Model, including the masses of the various elementary particles and the strengths of the basic forces. These parameters form a kind of configuration space for the laws of nature — a space called the *landscape of theories* (a term borrowed from population biology, where the space of genes is called the *fitness landscape*).

- *A mechanism of reproduction.* I adopt an old idea proposed to me by my postdoctoral mentor, Bryce DeWitt, which is that black holes lead to the births of new universes. This is a consequence of the hypothesis that quantum gravity does away with the singularities where time begins and ends — a hypothesis for which there is good theoretical evidence. Our universe has a lot of black holes, at least a billion billion of them, which suggests a very large population of progeny. We can suppose that our universe is itself part of a line of descent stretching far into the past.

- *Variation.* Natural selection works in part because genes mutate or recombine at random during reproduction, so that the genomes of offspring differ from that of either parent. Analogously, we can hypothesize that each time a new universe is created there is a small random change in the parameters of the laws. Thus we can mark on the landscape the point corresponding to the values of the parameters for that universe. The result is a vast and growing collection of points on the landscape, representing variations in the parameters of the laws across the multiverse.

- *Differences in fitness.* In population biology, the fitness of an individual is a measure of its reproductive success — that is, how many offspring it produces who thrive long enough to have children of their own. The fitness of a universe is then a measure of how many black holes it spawns. The number turns out to depend sensitively on the parameters. It's not easy to make a black hole; therefore many parameters lead to universes that have no black holes at all. A few parameters lead to universes that have lots of black holes. These universes occupy a very small region of the parameter space. We will assume that these highly fertile regions in the parameter space are islands surrounded by regions of much lower fertility.

- *Typicality.* We also assume that our own universe is a typical

member of the population of universes, as that population is
after many generations. Thus we can predict that any properties
shared by most universes are properties of our own.[4]

The power of natural selection as a methodology is such that strong
conclusions can be drawn from these minimal assumptions. The basic
consequence is that after many generations most universes have parameters within the highly fertile regions. It follows that if we change
the parameters of a typical universe, the result will most likely be a
universe that forms many fewer black holes. Since our universe is typical, this must be true of our universe as well.

This is a prediction that can be checked indirectly. We already know
that many ways of changing the parameters of the Standard Model result in universes without the long-lived stars needed to produce carbon and oxygen. And, remarkably, carbon and oxygen are necessary to
cool the gas clouds in which the massive stars that give rise to black
holes are formed. Other ways to change the parameters weaken the supernovas that not only lead to black holes but inject energy into the interstellar medium — energy that drives the collapse of the clouds, thus
forming new massive stars. We already know of at least eight ways to
slightly change the parameters of the Standard Model that would lead
to universes with fewer black holes.[5]

Cosmological natural selection thus offers a genuine explanation
for why the parameters of the Standard Model appear to be tuned for
a universe that is filled with long-lived stars that over time have en-
riched the universe with carbon, oxygen, and other elements needed
for the chemical complexity our universe is blessed with. The parameters whose values are thus to a greater or lesser extent explained include the masses of the proton, neutron, electron, and electron neutrino, and the strengths of the four forces. There's a bonus: While the
explanation involves maximizing the production of black holes, a consequence is to make the universe hospitable to life.

Moreover, the hypothesis of cosmological natural selection makes
several genuine predictions, which are falsifiable by currently doable
observations. One is that the most massive neutron stars cannot be

heavier than a certain limit. The idea here is that a supernova leaves behind the exploded star's central region. This core will collapse to either a neutron star or a black hole. Which of the two is produced depends on how much mass the core has; a neutron star can exist only if its mass is below a certain critical value. If cosmological natural selection is right, that critical value should be tuned as low as possible, because the lower it is, the more black holes are made.

It turns out that there are several possibilities for what neutron stars are made of. One possibility is just neutrons, in which case the critical mass would be rather high, between 2.5 and 2.9 times the mass of the sun. But another possibility is that a neutron star's center contains exotic particles called kaons. This would lower the critical mass compared with the neutrons-only model. The extent of that lowering, though, depends on the details of theoretical modeling; the various models give a critical mass somewhere between 1.6 and 2 times the solar mass.

If cosmological natural selection is right, we would expect that nature has taken advantage of the possibility of making kaons in the center of neutron stars to lower the critical mass. This could have been accomplished, it turns out, by tuning the mass of the kaon to be light enough; this can be done without affecting the rates of star formation by tuning the mass of the strange quark. When cosmological natural selection was first proposed, the heaviest neutron stars known had masses of less than 1.5 times that of the sun. But recently a neutron star has been observed that has a mass just under twice that of the sun. This would refute cosmological natural selection if the mass of the kaon-neutron stars is at the lower end of the theoretical range, but the theory just manages to fit if the right answer is the upper theoretical estimate, which is also twice the mass of the sun.

However, there is a less accurately measured neutron star whose mass is estimated to be as much as two and a half times that of the sun.[6] If that finding holds up under more precise measurements, cosmological natural selection will be falsified.[7]

Another prediction comes from thinking about a surprising feature of the early universe, which is its extreme regularity. The distribution of matter in the early universe is known, from observations of the CMB, to have varied only slightly from place to place. Why was this? Why did the universe not begin with large variations in density? If there were large variations in density, the highly dense regions would have collapsed right away to black holes. If the variations in density were large enough, these so-called primordial black holes would have filled the early universe, leading to a world with many more black holes than our own. This seems to falsify the prediction of cosmological natural selection, which is that there be no way to make a small change in the parameters of the laws of physics to make a universe with more black holes than our own.

Cosmologists describe the variations in the density of matter by a parameter called the scale of density fluctuations. This is not a parameter of the Standard Model of Particle Physics, but there are models of the early universe that do have adjustable parameters that can increase the density fluctuations, and it's fair to ask whether these are incompatible with cosmological natural selection. In most versions of inflation, there is a parameter that can be increased to raise the level of density fluctuations and thus flood the universe with primordial black holes. But in some of the simplest inflation models, raising this parameter shrinks the universe by sharply limiting the time over which the universe can inflate. The result is a much smaller universe, which, though filled with primordial black holes, has overall many fewer black holes than our own.[8] This means that cosmological natural selection is compatible only with a simple theory of inflation that cannot overproduce primordial black holes. If evidence is found that inflation happened in a way requiring a more complex theory, cosmological natural selection would be ruled out.[9] That there be no such evidence is hence a prediction of cosmological natural selection.

Of course the right theory of the very early universe may not be inflation, but this example serves to show that cosmological natural selection is vulnerable to disproof by any discovery of a mechanism acting in the early universe that might have produced many primordial black holes.[10]

Cosmological natural selection is inconceivable outside the context in which time is real. One reason is that all that need be claimed is that our universe has only a relative fitness advantage over universes differing by small changes in the parameters. This is a very weak condition. We needn't assume that the parameters of our universe are the largest possible; there very well might be other parameter choices leading to an even more fertile universe. All the scenario predicts is that they can't be reached by making a small change from the present values.

Thus the population of universes may be diverse, consisting of a variety of species, each relatively fertile compared with those that are slightly different. The mix of kinds of universes will continually change over time, as new ways to be fertile are discovered by trial and error. This is the way biology works. There are no maximally fit species that persist forever; rather, every era in the history of life is characterized by a different mix of species, all relatively fit. Life never reaches an equilibrium, or ideal state; it is ever evolving. Similarly, whatever laws are typical in the population of universes will change in time, as the population evolves. Were there a final state — in which, once reached, the mix of universes would stay the same — time would cease to matter, and we could say that a timeless equilibrium had been reached. But the natural-selection scenario does not assume or imply that. Time is always present in the scenario of cosmological natural selection.

Moreover, the scenario requires that time be universal as well as real. The population of universes evolves rapidly, growing each time each universe makes a black hole. If we are to deduce predictions from the theory, it must establish how many universes have such-and-such properties at each moment of time. This time must be meaningful not only throughout each universe but across the whole population. So we need a notion of time that gives us a picture of simultaneity within each universe and across that population.[11]

❖

Let's contrast this with the case of eternal inflation. The early universe is posited to inflate because the quantum fields accounting for its particles and forces are in a phase that produces a very large dark energy.

This makes the universe expand exponentially fast. Inflation typically stops when a bubble forms as a result of a phase transition. This is analogous to a bubble of water vapor appearing in a heated pot of water; the bubble contains the gas phase of water, which forms out of the liquid phase. In the cosmological scenario, the bubble contains a phase of the quantum fields that lacks the large dark energy, so its expansion slows and it becomes our universe.

What Vilenken and Linde noticed is that the surrounding medium, still containing the large dark energy, will continue to inflate rapidly. More bubbles form, which then become more universes, as our own did. They found that under certain conditions the process can go on forever, because the inflating medium never goes away, even as it produces an infinite number of bubble universes. If this scenario is correct, then our universe is one of an infinite number formed as bubbles in the eternally inflating medium.

In the simplest version, which I will presume for the purposes of our discussion, the laws that govern each bubble are chosen randomly from a landscape of possible laws.[12] In many discussions, this landscape is presumed to be given by various string theories, but any theory with variable parameters, including the Standard Model itself, will do.

In the simplest case, the proportions of bubbles that choose each law are constant, so as more and more bubble universes are produced, the probabilities for different laws to hold in the overall population remain the same. In such a simple scenario, time and dynamics play no role in how our universe's laws are specified among all the other (perhaps infinite) possibilities. The distribution of universes (that is, the probabilities for universes to have different laws or properties) thus reaches a kind of equilibrium and stays there forever. The scenario is in this sense timeless, which makes it a good case to contrast with cosmological natural selection.

Since the laws in each bubble are chosen randomly, universes with the finely tuned laws needed for life are exceedingly rare. So our universe is an atypical universe in the population of bubble universes.

To connect this scenario to observations of our universe, cosmologists must resort to the anthropic principle, which, as noted, states that we can live only in a universe whose laws and initial conditions create a world hospitable to life. The anthropic principle directs us to select the tiny fraction of hospitable universes from the vastly greater collection of lifeless worlds because we could find ourselves only in one of the former.

Remarkably, there are many commonalities in the lists of features that make a world hospitable to life and highly productive of black holes. So the two theories — cosmological natural selection and the anthropic principle — appear to explain some of the same fine-tunings of parameters of the Standard Model. But notice how different the explanations are. In cosmological natural selection, our world is a typical universe and most of the population will share the features that give a universe high fitness, whereas in the multiverse of eternal inflation, worlds like ours are extremely rare. In the first case we have a genuine explanation; in the latter, only a selection principle.

These different kinds of explanation differ in their ability to make genuine predictions for features of the universe not yet observed. As we saw, cosmological natural selection has already implied a few genuine predictions. But the scenarios invoking the anthropic principle as an explanation for our universe's laws and initial conditions have yet to yield a single falsifiable prediction for a currently doable experiment. I doubt they ever will.

Here's why. Consider any property of our universe you might want to explain. Either this property is necessary for intelligent life or it is not. If the former, that property is already explained by our existence, as it must hold in any of the very small fraction of universes with intelligent life. Now consider a second class of properties, those not necessary for intelligent life. Since the laws are chosen randomly in each bubble, these properties are randomly distributed in the population of universes. But since these properties have nothing to do with life, they will also be randomly distributed in the collection of living universes. So the theory makes no prediction for what we in our universe should observe about these properties.

The mass of the electron is a good example of the first kind of property; there's good evidence that conditions for life would deteriorate if the electron's mass differed much from its observed value.[13] A good example of the second kind of property is the mass of the top quark; as far as we know, it could vary within a large range without affecting the biofriendliness of our universe. Hence the anthropic principle cannot help us explain the observed value of its mass.

Eternal inflation does make one potentially testable prediction, which is that the curvature of space in every bubble universe is slightly negative. (A negative curvature space is warped like a saddle — in contrast to a positive curvature space, like a sphere.) If our universe was created in a bubble in an inflating multiverse, this must be true of it as well. That's a genuine prediction, but there are several problems regarding its testability. First, the negative curvature is very close to zero, and zero is hard to distinguish from a very small number, positive or negative. Indeed, the curvature vanishes within experimental error. Even with the better data expected from experiments in progress, it will be very hard to tell whether the curvature is exactly zero, slightly negative, or slightly positive. As in any scientific experiment, there will always be some uncertainty in the measurements. Given this uncertainty, it's unlikely that any observation will soon falsify the prediction.

Even if we did manage to verify that the spatial curvature of our universe is slightly negative, this does not provide evidence that our universe is one of a vast multiverse. There are lots of cosmological models and scenarios consistent with the curvature being slightly negative; one of them is that our universe is unique and is simply a solution of Einstein's equations with negative curvature. Such solutions exist and do not require inflation to justify them. Another is that inflation produced just one universe. And no observation can confirm a hypothesis about the properties of an alleged collection of other universes that do not in any way affect our own.

◆

The eternal-inflation scenario requires a set of possible theories, and this can be supplied by the huge number of possible string theories.

That there was a large landscape of possible string theories was evident from Strominger's aforementioned 1986 paper, but the situation became a crisis impossible to ignore when, in 2003, evidence was discovered for the existence of an astronomical number of string theories with small positive values of the cosmological constant.[14] The number was roughly estimated at 10^{500}. However, at least so far, the number, though enormously large, is still finite. Then in 2005, MIT physicist Washington Taylor and colleagues were able to construct evidence for an infinite number of string theories with small *negative* cosmological constants.[15]

This has an interesting consequence, which was pointed out by the South African physicist George F. R. Ellis.[16] If there really are an infinite number of string theories with small negative values of the cosmological constant but only a finite number with a small positive cosmological constant, then we should predict the cosmological constant to be small and negative. If the actual value is randomly distributed among universes in the multiverse, then we are infinitely more likely to be living in a universe with a negative value than in one with a positive value, because there are infinitely more of the former than the latter. This would be a genuine prediction of string theory, and such things are rare. Taken at face value, it indicates that the theory is wrong, since the measured value of the cosmological constant is positive.

Some string theorists have cautioned that there is still much to be discovered about constructing string theories, so evidence may yet be discovered for an infinite number of string theories with positive values of the cosmological constant. Another response has been to invoke the anthropic principle to argue that the universes with negative cosmological constant values described by Taylor and his colleagues should be ruled out because they are inhospitable to life.[17] However, all we need for the infinitude of negative cosmological constant universes to dominate over the finite number of universes with a positive cosmological constant is that any finite fraction of the former contain life.

The problem with anthropic cosmology is that you can always manipulate the assumptions when you're dealing with theoretical entities such as other universes that are in principle unobservable.[18] We cannot verify the hypothesis that there are a vast or infinite number of other universes, nor can we count how various properties are distributed among them. We can argue about whether universes different from ours might or might not have life, but we cannot check our arguments with observations.

A telling difference between anthropic theories and cosmological natural selection is how the two address the puzzling problem of the cosmological constant. As noted, this important constant of physics has been measured and has a tiny but positive value: in units of the Planck scale, 10^{-120}. The mystery is why it is so tiny. One relevant fact is that if we increase the cosmological constant from its observed value, keeping all the other constants in physics and cosmology fixed, we soon reach a value where the universe expands so fast that galaxies never form. Let's call this the *critical value*. It is around twenty times the observed value.

Why is this relevant? I'll start with a fallacious argument that goes like this:

(1) Galaxies are necessary for life. Otherwise stars would not form, and without stars there is no carbon and no energy to promote the emergence of complex structures, including life, on the surfaces of planets.
(2) The universe is full of galaxies.
(3) But the cosmological constant has to be smaller than the critical value if galaxies are to form.
(4) Hence, the anthropic principle predicts that the cosmological constant must be smaller than the critical value.

Can you see the fallacy? Point no. 1 is true, but it plays no role in the logic of the argument. The real argument starts with point no. 2. The fact that the universe is filled with galaxies is evident from observations; it is irrelevant whether or not life would be possible without them. So the first point can be dropped from the argument without weakening the conclusion. But point no. 1 is the only place life is mentioned, so once it's dropped, the anthropic principle plays no role. The correct conclusion is:

(4) Hence, the observed fact that the universe is full of galaxies implies that the cosmological constant must be smaller than the critical value.

One way to tell that the argument is fallacious is to ask how we would respond if the cosmological constant turned out to be above the critical value. We would not challenge assertion no. 1, which is in any case irrelevant. We would not challenge no. 2, which is a statement of fact. We could challenge only no. 3, which is a theoretical statement. Maybe our computation of the critical value is wrong.

In 1987, Steven Weinberg proposed an ingenious explanation for the small value of the cosmological constant which is not subject to this fallacy but still uses the anthropic principle.[19] It goes like this: Hypothesize that our universe is one of a vast multiverse in which the values of the cosmological constant are randomly distributed between zero and one.[20] Since we require galaxies to live, we must live in one of the universes with a cosmological constant below the critical value. But we could live in any one of those. Therefore our situation is as if the cosmological constant were pulled out of a hat, randomly chosen to be some number between zero and the critical value. This implies that it is improbable that the value of our cosmological constant is a lot smaller than the critical value, because only a tiny fraction of the numbers in the proverbial hat will be that small. We should expect the cosmological constant in our universe to be of the same order of magnitude as the critical value, because there are many more numbers roughly that size than there are numbers that are much smaller.

On this basis, Weinberg predicted that the cosmological constant should be below, but within an order of magnitude of, the critical value. And remarkably, when the cosmological constant was measured ten years later,[21] it was found to be about 5 percent of the critical value. By the reasoning just given, that would happen about one in twenty times we picked a number from a hat. This is not so unlikely, lots of things happen in the world that have a one-in-twenty chance. So some cosmologists argue that the success of Weinberg's prediction can be taken as evidence in favor of the hypothesis it was based on — that we live in a multiverse.

One problem with that conclusion is that the critical value referred to is the one above which galaxies would not form *if the cosmological constant were the only parameter that varied.* But theories of the early universe have other parameters that can vary. If we vary some of those while we vary the cosmological constant, the argument loses its force.[22]

Let's look at one case, in which we vary the size of the density fluctuations, which, as we discussed earlier in this chapter, determine how evenly the matter in the early universe was distributed. These are relevant because if they were bigger, the cosmological constant could be far above the critical value and galaxies would still form in the very dense regions created by the fluctuations. There is still a critical value for the cosmological constant, but it goes up as the size of the density fluctuations goes up.

So you can rerun the argument, letting the cosmological constant and the fluctuation size both vary over the population of universes. Now you pull two numbers out of the hat for each universe, one for the cosmological constant, the second for the size of the density fluctuations. We choose these numbers randomly, within the range in which galaxies form.[23] It turns out that the probability of randomly getting both numbers to be as small as they are observed to be is now down from 1 chance in 20 to a few parts in 100,000.[24]

The problem is that because we don't observe any other universes, it is impossible to know which constants vary over the hypothetical multiverse. If we assume that the right story is that only the cosmological constant can vary over the multiverse, Weinberg's argument does well. If we assume that the right story is instead that both the cosmological constant and the fluctuation size vary, the argument does less well. In the absence of any independent evidence as to which, if any, of these hypotheses are true, the argument leads to no conclusion.

So the claim that Weinberg's argument correctly predicted the rough value of the cosmological constant fails, because of a subtler fallacy than the one discussed above. This fallacy, which is known to specialists in probability theory, arises whenever you take advantage of the freedom to arbitrarily choose a probability distribution that describes unobservable entities and so cannot be checked independently. Weinberg's original argument has no logical force, because you could reach a different conclusion by making a different assumption about unobservable entities.[25]

Cosmological natural selection does better at explaining the same evidence, because it provides a reason to fix both the fluctuation size and the cosmological constant. Recall that in some simple models of inflation, the fluctuation size is strongly anticorrelated with the size of the universe; that is, the smaller the fluctuation size, the bigger the universe, and hence (everything else being equal) the more black holes. So the fluctuation size should be near the lower limit of the range required for galaxies to form. This in turn implies a small value for the critical value of the cosmological constant consistent with galaxy formation. Cosmological natural selection, together with the simple model of inflation, predicts that both the fluctuation size and the cosmological constant should be small. This prediction is not arbitrary, and it fits the evidence well.

The anthropic principle, however, is compatible with a much smaller universe, because a single galaxy is probably enough to give rise to intelligent life. Observations suggest that a high proportion of stars have planets, so a galaxy's worth of planets should be enough to ensure that at least one of them has life. Adding more galaxies will not increase the probability that life will arise

An anthropic enthusiast might counter that the anthropic principle can be saved by modifying it to say that we are more likely to inhabit a universe with a high number of planets that harbor life. This gives a reason to prefer universes as large as possible, and this implies a low value for both the density fluctuations and the cosmological constant.

Something funny must be going on, because we are apparently altering the prediction of a theory without actually changing any facts. The two versions of the anthropic principle do not differ in any assertion about the actual multiverse but only in how we select the universes we feel we ought to consider from a much larger population of inhospitable universes.

"Wait a minute," the anthropic enthusiast might reply. "A civilization in a multiverse is more likely to find itself in a universe with many civilizations and hence many galaxies than a universe with just one galaxy." This seems at first to be a plausible argument, but we have to counter, "How do you know?" There could be many more small universes in the multiverse than large universes, so that a randomly chosen civilization is more likely to be in a small universe. Which scenario is correct depends on the relative distribution of large and small universes in the multiverse, but this feature cannot be independently verified. Theorists could probably produce different models, favoring different distributions of sizes of universes, but the fact that you can adjust unobservable features of your scenario to enable you to pick one that fits your hypothesis better does not constitute evidence for that scenario.

In cosmological natural selection, however, our universe is a typical member of the population of universes, and there is no room to insert an adjustable principle to select out atypical cases.

Notice that the argument is not about creating universes in black holes versus creating them as bubbles during inflation. It is about the role of time and dynamics in the logic by which the scenarios explain known features of the universe and predict new ones. An inflationary model could employ time and long chains of descent — bubbles within bubbles within bubbles — in such a way as to avoid dependence on the anthropic principle and enjoy the advantages of cosmological natural selection.

The point is not only that the theory that posits continual evolution over time does better than the timeless theory in fitting the observational evidence. The point is also that the theory calling for evolution makes a clean prediction, whereas the predictions of the anthropic argument are adjustable depending on how we want the argument to run. Contrary to what we might first have thought, hypotheses based on the idea that the laws of nature evolve over time are more vulnerable to falsification than are timeless cosmological scenarios. And if an idea is not vulnerable to falsification, it is not science.

12

Quantum Mechanics and the Liberation of the Atom

WE HAVE SEEN THAT the reality of time is the key to addressing the mystery of what selects the laws of physics. It does so by supporting the hypothesis that those laws evolve. Taking time as fundamental may also help resolve another great puzzle of physics — that of making sense of quantum mechanics. Time's reality allows a new formulation of quantum theory that can also illuminate how laws evolve in time.

Quantum mechanics is the most successful physical theory yet invented. Almost none of the digital, chemical, and medical technologies we rely on would exist were it not for quantum physics. Yet there is strong reason to believe that the theory is incomplete.

Certainly quantum mechanics is a challenge to our attempts to comprehend the world. Since its invention in the 1920s, physicists have concocted bizarre scenarios to make sense of the puzzles of quantum theory. Cats that are both alive and dead, an infinitude of simultaneously existing universes, reality that depends on what's measured or who's observing, particles that signal each other across vast distances at speeds exceeding that of light — these are some of the imaginative ideas proposed as resolutions of the mysteries of the subatomic world.

All these strategies arise as a response to the fact that quantum mechanics gives no physical picture of what's going on in an individual experiment. This is not disputed. The axioms of quantum mechanics include the statement that it gives only statistical predictions of the outcomes of experiments.

Einstein made the case long ago that quantum mechanics is incomplete because it fails to give a precise description of what goes on in an individual experiment. What exactly is the electron doing when it jumps from one energy state to another? How do particles too far away from each other to instantaneously communicate do so? How do they appear to be in two places at once? Quantum mechanics gives no answers. Nevertheless it is extraordinarily useful, in part because it provides physics with a language and a framework for organizing vast amounts of empirical data. If it fails to show us what actually happens at the subatomic level, it does give us an algorithm for predicting the probabilities for different experimental outcomes. And so far, the algorithm works.

Can a theory be successful as a generator of predictions and still be off the mark, in the sense that future theories may overturn the assumptions it makes about the world? This has happened several times in the history of science. The assumptions underlying Newton's laws of motion were overturned by relativity and quantum theory. Ptolemy's model of the solar system served us well for more than a millennium, yet it was based on ideas that are wildly wrong. It would seem that effectiveness is no guarantee of truth.

I have come to believe that quantum mechanics will suffer the same fate as the great theories of Ptolemy and Newton. Perhaps we can't make sense of it simply because it isn't true. It is instead likely to be an approximation to a deeper theory that will be easier to make sense of. That deeper theory is the unknown cosmological theory all the arguments of this book point toward. The key is, again, the reality of time.

Quantum mechanics is a problematic theory for three closely related reasons. The first is its failure to give a physical picture of what goes on in an individual process or experiment; unlike previous physical theories, the formalism we use in quantum mechanics cannot be read as showing us what's happening moment by moment in time. Second, in most cases it fails to predict the *precise* outcome of an experiment; rather than telling us what will happen, it gives only probabilities for the various things that might happen.

The third and most problematic feature of quantum mechanics is that notions of measurement, observation, or information are necessary to express the theory. These must be regarded as primitive notions; they cannot be explained in terms of fundamental quantum processes. Quantum mechanics is not a theory so much as a method for coding how experimenters interrogate microscopic systems. Neither the measuring instruments we use to interact with a quantum system nor the clock we use to measure time can be described in the language of quantum mechanics — nor can we, as observers, be so described. This suggests that to make a valid cosmological theory we will have to give up quantum mechanics and replace it with a theory that can be extended to the whole universe, including ourselves as observers and our measuring instruments and clocks.[1]

As we seek that theory, we must keep in mind three clues about nature that experiment has revealed are integral to quantum physics: *incompatible questions, entanglement,* and *nonlocality.*

Any system will have a list of properties, such as position and momentum[2] for particles or color and heel height for shoes. Associated with each property is a question that can be asked: Where is the particle now? What color is her shoe? It is the role of experiment to interrogate the system to get answers to these questions. If you want to describe a system in classical physics completely, you answer all the questions, and this gives you all the properties. But in quantum physics, the set-up you need for asking one question may render other questions unanswerable.

For example, you can ask what a particle's position is, or you can ask what its momentum is, but you cannot ask for both at once. This is what Niels Bohr called *complementarity,* and it is also what physicists mean when we talk of *noncommuting variables.* If there is quantum fashion, then shoe color and heel height might be incompatible properties. This is very different from classical physics, where you don't have to choose what property to measure and what to leave out. The crucial question is whether the choice the experimenter must make influences the reality of the system she's studying.

Entanglement, too, is a purely quantum phenomenon, according to which pairs of quantum systems can share properties while each system remains individually indefinite. That is, you can ask a question about a relationship between the pair that has a definite answer, whereas the answer to any related question about the individuals does not. Consider a pair of quantum shoes. They can have a property called *contrary,* according to which any question asked of both of them will give opposite answers. If you ask both of them what color they are and the left shoe answers "white," then the right shoe will answer "black," and vice versa. If you ask about heel height, then if the left shoe's heel is high, the right shoe's heel will be low, and vice versa. If you ask only about the height of the left shoe's heel, the answer will be either "high" or "low" with a 50-percent probability. Similarly, with regard to one shoe's color, the answer will be either "black" or "white" with a 50-percent probability. In fact, if the pair of quantum shoes has the property *contrary,* then any question addressed to one shoe alone will evoke random answers and any question addressed to both will evoke contrary answers.

In classical physics, any property of a pair of particles is reducible to a description of properties of each. Entanglement shows that this is not true for quantum systems. The importance for our discussion is that you can create, through entanglement, novel properties in nature. If you entangle two quantum systems of a kind that have never before interacted with each other, by preparing them with a property like *contrary* you create a property that has never before existed in nature.

Entangled pairs are created by bringing two subatomic particles together and having them interact. Once entangled, they stay entangled, even if they separate and move a great distance away from each other.

As long as neither one interacts with another system, they continue to share entangled properties, such as *contrary*. This gives rise to the third and most startling clue about nature at the quantum level, which is *nonlocality*.

Let's entangle a pair of shoes with the property *contrary* in Montreal and send the left shoe to Barcelona and the right shoe to Tokyo. Experimenters in Barcelona choose to measure the left shoe's color. This choice appears to instantaneously affect the color of the right shoe in Tokyo. This is because once the Barcelona lab has observed the color of their shoe, they can correctly predict that the Tokyo shoe has the opposite color.

In the 20th century, we became accustomed to physical interactions having a property called locality, which meant that if information was to be transferred from one place to another, it had to travel by means of a particle or a wave. Because of special relativity, any influence was presumed to travel at the speed of light or slower. Quantum physics appears to violate this central tenet of special relativity.

The nonlocal effects in quantum theory are real, but they are subtle and cannot be used to send information between Barcelona and Tokyo. The reason is that whatever property the experimenters in Tokyo choose to measure, the result will appear to them to be random. They will see their shoe to be black or white equally often. It is only when they learn what color was seen in Barcelona that they will realize that the pairs are opposite in color. But to understand this requires that information be transmitted between Barcelona and Tokyo — i.e., at the speed of light or less.

The question that remains, however, is how the correlations are established between the shoes in Tokyo and Barcelona, so that when the experimenters there each open their box and take out their shoe, the colors are always opposite. One might think that whoever packed the boxes in Montreal made sure to put one color in the box headed to Tokyo and the opposite color in the box headed to Barcelona. However, it can be proved by a combination of theoretical arguments and experimental results that this is exactly what does not happen. Instead, the correlations are established somehow at the moment the boxes are opened in Tokyo and Barcelona.

Let's suppose we have a big box full of pairs of shoes and we've entangled each pair with the property *contrary.* We ship all the left shoes to Barcelona and all the right shoes to Tokyo. Let the experimenters in each city choose randomly what property of each individual shoe they measure and keep track of the results. They send their choices and results back to the factory in Montreal, where they are compared. It turns out that the only way to make sense of the joint results is to assume that there are nonlocal effects, by which the properties of one shoe of each pair are influenced by the choices made about what to measure on the other shoe. This is the content of a theorem proved in 1964 by Irish physicist John Stewart Bell and demonstrated by a related set of clever experiments.

These features and issues have been the focus of a great deal of attention in the nine decades since quantum mechanics was formulated. Many approaches to a greater understanding of it have been proposed. I believe now that they all miss the mark, and that the strange features of quantum theory arise because it is a truncation of a cosmological theory—a truncation applicable to small subsystems of the universe. By embracing the reality of time, we open a path to understanding quantum theory that illuminates its mysteries and may well resolve them.

◈

I further believe that the reality of time makes a new formulation of quantum mechanics possible.[3] This formulation is new and speculative. It has not yet led to any precise experimental predictions, let alone experimental tests, so I cannot claim that it's correct. What it does do is offer a radically different perspective on the nature of physical laws, realizing in a new and surprising way the idea of laws evolving in time. And it will likely be testable, as we will see shortly.

But can we really give up the idea of timeless laws of nature without losing the power of physics to explain so much of the world around us? We're used to thinking that the laws are deterministic. Among other consequences of determinism is that there cannot be anything genuinely new in the universe — that everything that happens is the rearrangement of elementary particles with unchanging properties by unchanging laws.

There are certainly countless situations in which the future can reliably be expected to mirror the past. When we do an experiment that we have carried out many times before and in which we have always gotten the same result, we can reliably expect that result in the future. (Even if the results are sometimes one thing and sometimes another, the proportions of each outcome will hold in future measurements.) We can expect that the next time we throw a ball, it will travel along a parabola, which is what has happened every time we've done this in the past. Usually we say that this is because the motion is determined by a timeless law of nature, which, being timeless, will act in the future just as it has acted in the past. So timeless law precludes genuine novelty.

But is the assumption that a timeless law acts really necessary to explain that the present mirrors the past? We need the notion of a law only in cases in which a process or experiment has been repeated many times. But to explain these cases, we actually need a lot less than a timeless law. We could get by with something much weaker — say, a principle stating that repeated measurements yield the same outcomes. Not because they are following a law but because the only law is a *principle of precedence.* Such a principle would explain all the instances in which determinism by laws work but without forbidding new measurements to yield new outcomes, not predictable from knowledge of the past. There could be at least a small degree of freedom in the evolution of novel states without contradicting the application of laws to circumstances that were repeatedly produced in the past. Common law in the Anglo-Saxon tradition operates by a principle of precedence, whereby judges are constrained to rule as judges have in the past, when presented with similar cases. What I want to suggest is that something like this might well be operating in nature.

Once I formulated this idea, I was astonished to learn that here, as well, I have been preceded by Charles Sanders Peirce, who spoke of the laws of nature as habits developed over time:

[A]ll things have a tendency to take habits. For atoms and their parts, molecules and groups of molecules, and in short every conceivable real object, there is a greater probability of acting as on a former like occasion than otherwise. This tendency itself constitutes a regularity, and is continually on the increase. In looking back into the past we are looking toward periods when it was a less and less decided tendency.[4]

This principle would become crucial in genuinely novel cases. For if nature is really operating according to a principle of precedence rather than timeless law, then when there are no precedents there will be no prediction for how a system will behave. If we produce a truly novel system, its response to measurement will not be predictable from any information we already have. Only once we have produced many copies of that system does the principle of precedence take over. From then on, the system's behavior is predictable.

If nature is like this, then the future is genuinely open. We would still have the benefit of reliable laws in cases with ample precedent, but without the stranglehold of determinism.

It is fair to say that classical mechanics precludes the existence of genuine novelty, because all that happens is the motion of particles according to fixed laws. But quantum physics is different, in two ways that enable us to replace timeless laws by a principle of precedence.

First, as we have seen, entanglement can produce genuinely novel properties. You can test a pair of particles for an entangled property like *contrary* that is a property of neither particle separately. Second, there appears to be an element of genuine randomness in the response of quantum systems to their environments. Even if you know everything about the past of a quantum system, you cannot reliably predict what it will do when one of its properties is measured.

These two features of quantum systems let us replace the postulation of timeless laws with the hypothesis that a principle of precedence acts in nature to ensure that the future resembles the past. This principle is sufficient to uphold determinism where it's needed but implies that nature, when faced with new properties, can evolve new laws to apply to them.

Here's a simple illustration of the operation of the principle of precedence in quantum physics: Consider a quantum process in which a system is prepared and then measured, and assume that this process has occurred many times in the past. This gives you a collection of past outcomes of the measurement: X many times the system said yes to a question, and Y many times it said no. The outcome of any future instance of that process is then picked randomly from the collection of the outcomes of past cases. Now suppose that there's no precedent, because this system has been prepared with a definite value of a genuinely novel property. Then the outcome of the measurement will be free, in the sense that it is not determined by anything in the past.

Does this idea mean that nature is really free to choose the outcome of an experiment? There's a certain sense in which quantum systems are already known to have an element of freedom—a sense illuminated by a recent theorem invented and proved by John Conway and Simon Kochen, two Princeton mathematicians. I don't much like the name they gave their result, but it's catchy and has duly caught on: the *free-will theorem*.[5] The theorem applies to the case of two atoms (or other quantum systems) that become entangled and then separate, after which a property of each is measured. The theorem says: Suppose there is a sense in which the two experimenters are free to choose which measurement they make on their atoms. Then the response of the atoms to the measurement is free in the same sense.

This need have nothing to do with the slippery concept of free will. If we assert that the experimenters are free to choose what measurement they make, we mean that their choice is not determined by their past history. No amount of knowledge about the past of the experimenters and their world will let us predict their choice. Then the atoms are also free, in the sense that no amount of information about the past enables us to predict the outcome of a measurement of one of their properties.[6]

I find it marvelous to imagine that an elementary particle is truly free, even in this narrow sense. It implies that there's no reason for what an electron chooses to do when we measure it — and thus that there's more to how any small system unfolds than could be captured in any deterministic or algorithmic framework. This is both thrilling and scary, because the idea that choices atoms make are truly free (i.e., uncaused) fails to satisfy the demand for sufficient reason — for an answer to every question we might ask of nature.

Can we quantify how much freedom nature has if quantum mechanics is correct? We know that classical mechanics has no such freedom, because it describes a deterministic world whose future can be completely predicted from knowledge of the past. Statistics and probabilities can play a role in the description of the classical world, but they only reflect our ignorance. No freedom is granted, because we can always learn enough to make definite predictions.

Conway and Kochen's theorem suggests that quantum systems have a degree of real freedom, but could there be a kind of physics according to which nature has even more freedom? I asked myself this question, and it was not too hard to answer. To do so, I relied on recent work in quantum foundations that gave me a precise definition of how much freedom a quantum system may have.

Around 2000, Lucien Hardy, then at Oxford University but shortly to move to the Perimeter Institute for Theoretical Physics, conceived of a general class of theories that predict probabilities for outcomes of measurement. These included not just classical and quantum mechanics but many other theories as well. Hardy required only that the theories make consistent use of the notion of probability and behave reasonably when applied either to an isolated system or to two or more systems in combination. These requirements are expressed in a short list of assumptions, or axioms, which Hardy calls "reasonable axioms."[7] They have been developed and modified by subsequent theorists. I was able to use an elaboration of Hardy's axioms invented by Lluís Masanes and Markus Müller[8] to make a precise statement about how much freedom a theory has.

The amount of freedom is expressed by how much information you need about a system to be able to make predictions about its future.

This information can be gained by preparing many identical copies of the system and asking different questions of each. The predictions this interrogation enables us to make may still be probabilistic, but they are the best possible predictions, in the sense that no further observations of the system would improve their precision. For each system Hardy studied, there's a certain finite amount of information you need in order to best ascertain what the system will do when faced with any possible measurement. The more things you need to measure about a system before you can make the best possible predictions, the more freedom it has.

To see how much freedom this implies, we should compare the amount of information needed to make predictions to some measure of the size of the system. One useful measure is the number of answers the system can give to an answer posed in an experiment. In the simplest case, there are only two choices: If you ask about the color of a quantum shoe, it can be either white or black. If you ask about heel height, it can be either high or low.

What I showed is that quantum mechanics maximizes the amount of information you need per choice. That is, quantum mechanics describes a universe in which you can make probabilistic predictions of how systems behave, but in which those systems have as much freedom from determinism as any physical system described by probabilities can have. So in the sense that quantum systems are free, they are maximally free. Combining the principle of precedence with this *principle of maximal freedom,* you get a new formulation of quantum physics. This formulation cannot be expressed outside a framework in which time is real, because it makes essential use of the distinction between past and future. So we can abandon the idea that there are timeless and deterministic laws of nature without losing any of the explanatory power of physics.

The result that quantum systems maximize their freedom was an almost trivial step, given the previous work of Hardy, Masanes, and Müller. The new perspective I brought to the problem was the reality of time.

The first reaction of some friends and colleagues when I explained this idea was laughter. Certainly there are details that remain to be filled in, such as how precedence builds up from the freedom of the first case, through the next few cases, and on to established cases with many precedents.[9] But beyond the details, the proposal for a principle of precedence does have an element of implausibility about it. How does a system recognize all its precedents? By what mechanism does a system pick out a random element in the collection of its precedents? This would seem to require a new kind of interaction, whereby a physical system can interact with copies of itself in the past.

The principle doesn't say how this takes place; in this respect it's no better than the usual formulation of quantum mechanics. In the older formulation, measurement is a primitive notion; in the present formulation, being a quantum system of the same kind (that is, prepared and transformed the same way) is a primitive notion. But one might have asked similar questions about the idea of a timeless law of nature acting to cause motion and change. How does an electron "know" it is an electron, so that the Dirac equation rather than another equation applies to it? How does a quark "know" which kind of quark it is and what its mass should be? How does a timeless entity, such as a law of nature, reach somehow inside time to act on every single electron?

We're accustomed to the idea of timeless laws of nature acting inside time, and we no longer find it strange. But step back far enough, and you can see that it rests on some big metaphysical presumptions that are far from obvious. The principle of precedence also relies on metaphysical presumptions, but these are less familiar to us than those that allow us to believe in timeless laws of nature.

If the metaphysics implied by the principle of precedence is novel, in my view it is far more parsimonious than some of the current fantastic approaches to quantum theory — such as that our reality is one of an infinite set of simultaneously existing worlds. When it comes to quantum theory, you have to embrace some very strange notions. But we're free to pick our own strange notions — at least until experiment tells us one approach to quantum theory is superior to the rest.

I'm willing to bet that the principle of precedence will generate new ideas for experiments whose results might point us to a physics beyond quantum mechanics.

You might object that quantum mechanics already provides predictions for how a novel property will behave. Does this new idea contradict those predictions? It does, and this is the most likely reason it might fail. Suppose we produce, in a quantum computer, a new kind of entangled state never before produced in nature. In conventional quantum theory, you could compute how this entangled system will behave when it's measured. The principle of precedence that I'm proposing suggests that these predictions might not be borne out by experiment. This is equivalent to suggesting that new kinds of entangled states give rise to new interactions in nature, or to context-dependent changes in existing interactions. Such novel interactions have never been observed, nor has a context dependence of interactions, and skepticism is in order.

But seldom in our history has human ingenuity led to the creation of new kinds of entangled states. We're just learning to do this, and if this new hypothesis is right, the outcomes of experiments with quantum computers may be surprising. At the least, it's probably vulnerable to falsification by experiments with quantum devices that produce novel entangled states. It contradicts a fundamental tenet of reductionism, according to which the future of compound systems, no matter how complex, can be predicted given only a knowledge of the forces existing between pairs of elementary particles. But the violations of reductionism involved are rare and mild, so I would urge that it is reasonable to let experiment decide.

This new understanding of quantum physics realizes two of the criteria for a cosmological theory. It satisfies the demand for *explanatory closure* (albeit in a restricted form, which allows for genuine freedom in novel cases). The principle of precedence says that what determines the outcome of future measurements is the collection of past cases. These cases were real, so we have only an effect of things that were real in the past on things that will be real in the future. Clearly, it also satisfies the criterion that laws evolve, and it does so provocatively, by proposing that unprecedented measurements are not governed by any prior law. As outcomes accumulate, precedent is established; only when sufficient precedence has been established are future outcomes law-like.

As new states arise in nature, new laws evolve to guide them — which suggests that the fundamental interactions we observe and describe with the Standard Model of Particle Physics resulted from the "locking in" of new laws when the states corresponding to electrons, quarks, and their relatives first emerged as the universe cooled shortly after the Big Bang.

Something this new proposal does *not* do is satisfy the principle of sufficient reason. To the extent that quantum systems really are free — in the sense that individual outcomes are undetermined — the principle of sufficient reason is thwarted, for then there's no rational reason for the outcome of an individual experiment. There simply is no reason for when a radioactive nuclei decays, or for the exact outcomes of any of the other cases for which quantum mechanics gives only probabilistic predictions.

Whatever the fate of this new idea — and, as with any speculative new idea, we must expect that it may fail — we can see the fruitfulness of the hypothesis of the reality of time. The reality of time is not just a metaphysical speculation; it is a hypothesis capable of inspiring new ideas and driving a robust research program.

13

The Battle Between Relativity and the Quantum

THE PRINCIPLE OF SUFFICIENT reason is central to the program of extending physics to the scale of the universe as a whole, because it sets a goal of discovering a rational reason for every choice that nature makes. The apparently free, uncaused behavior of individual quantum systems presents a formidable challenge to that principle.

Can the demand for sufficient reason be satisfied even in quantum physics? This depends on whether quantum mechanics can be extended to the universe as a whole and give the most fundamental description of nature possible or is only an approximation to a very different cosmological theory. If we can extend quantum theory to the universe as a whole, then the free-will theorem applies at the cosmological scale. Since we assume there is no theory more fundamental, it implies that nature is truly free. The freedom of quantum systems at the cosmological scale would imply a limit to the principle of sufficient reason, because no rational or sufficient reason could be given for the myriad of free choices quantum systems make.

But in proposing this extension of quantum mechanics, we commit the cosmological fallacy, wrenching a theory beyond the limited domain where it can be compared to experiment. A more cautious response would be to explore the hypothesis that quantum physics is an approximation, valid only for small subsystems. The missing information needed to determine what a quantum system will do might still be present somewhere out in the universe, so that it comes into play when we embed a quantum description of a small subsystem into a theory of the universe as a whole.

Could there be a deterministic cosmological theory that gives rise to quantum physics whenever we isolate a subsystem and ignore the rest? The answer is yes, but, as we are about to see, it comes at a high cost.

According to such a theory, the probabilities of quantum theory are due only to our ignorance of the whole universe, and the probabilities give way to definite outcomes at the level of the universe as a whole. The quantum uncertainties originate when the cosmological theory is truncated to describe a small part of the universe.

Such a theory has been called a *hidden variables* theory, because the quantum uncertainties are resolved by information about the universe which is hidden to the experimenter working on an isolated quantum system. Theories of this kind have been proposed and give predictions for quantum phenomena which agree with those of quantum physics. So we know that, at least in principle, this kind of resolution of the problems of quantum mechanics is possible. Moreover, if determinism is restored by extending quantum theory to a theory of the whole universe, then the hidden variables have to do not with a more precise description of the individual quantum system but with the relationship of that system to the rest of the universe. We can then call them *relational hidden variables*.

According to the principle of maximal freedom, described in the last chapter, quantum theory is the probabilistic theory in which intrinsic uncertainties are as large as possible. Another way to put this is that the information we would need about an atom to restore determinism, which is coded in relations of that atom to the universe as a whole, is maximal. That is, the properties of each particle in the universe are maximally tied up in hidden relations to the universe as a whole. Thus the problem of making sense of quantum theory is central to the search for the new cosmological theory the other arguments of this book point to.

Here's the price of admission: It means giving up the relativity of simultaneity and going back to a picture of the world in which an absolute definition of simultaneity holds throughout the universe.

We have to proceed cautiously here — we don't want to contradict the successes of relativity theory. Among these are a successful marriage of special relativity with quantum theory called quantum field theory. It is the basis of the Standard Model of Particle Physics and makes a great many precise predictions, which have been upheld by the results of many experiments.

Quantum field theory is, however, not without its problems. Among them are that a tricky game must be played with infinite quantities before any predictions can be extracted. Moreover, quantum field theory inherits all the conceptual problems of quantum theory and offers nothing new toward their resolution. These old problems of quantum theory, together with the new issues of infinities, suggest that quantum field theory, too, is an approximation to a deeper, more unified theory.

Thus, despite the successes of quantum field theory, many physicists, beginning with Einstein, have wanted to go beyond it to a deeper theory that gives a complete description of each individual experiment — which, as we have seen, no quantum theory does. Their searches have consistently found an irreconcilable conflict between quantum physics and special relativity. This conflict is something we need to understand as we contemplate the rebirth of time in physics.

◈

There is a tradition — starting with Niels Bohr — of claiming that quantum theory's failure to give a picture of what goes on in an individual experiment is one of its virtues rather than a defect. As noted in chapter 7, Bohr argued skillfully that the goal of physics is not to give such a picture but instead to create a language in which we can talk to each other about how we set up experiments on atomic systems and what the results are.

I find Bohr's writings fascinating but unconvincing. I feel the same way about some contemporary theorists, who argue that quantum mechanics is not "about" the physical world but about the *information* we have about the physical world. These theorists argue that the quantum state does not correspond to any physical reality; rather, it just codes the information we observers can have about a system. These are smart people, and I enjoy arguing with them, but I fear they're selling science short. If quantum mechanics is only an algorithm for predicting probabilities, can't we do better? After all, something *is* going on in an individual experiment. Something, and only that something, is the reality we call an electron or a photon. Shouldn't we be able to capture the essence of the individual electron in a conceptual language and a mathematical framework? Perhaps there is no principle guaranteeing that the reality of each subatomic process in nature should be comprehensible to human beings and expressible by us in language or mathematics. But shouldn't we at least make the attempt? So I side with Einstein. I believe there is an objective physical reality and that something describable happens as an electron jumps from one energy level of an atom into another. I then seek a theory to give this description.

The first hidden-variables theory was presented by Prince Louis de Broglie in 1927 at an iconic gathering of quantum physicists called the Fifth Solvay Conference, shortly after quantum mechanics was put in final form.[1] It was inspired by the duality between wave and particle that Einstein had suggested, which we discussed in chapter 7. De Broglie's theory resolved the conundrum of wave and particle in a way that is simplicity itself. He posited that there is a real particle and a real wave. Both have material existence. Earlier, in his 1924 PhD thesis, he had posited that the wave/particle duality is universal, so that particles, such as electrons, are also waves. In de Broglie's 1927 paper, these waves propagate like water waves, interfering and diffracting. The particle follows the wave. In addition to the usual forces — electricity, magnetism, and gravity — the particle is pulled by a force called the quantum force. This force pulls the particle toward the wave crest; hence, on average, the particle is more likely to be found there, but the connection is probabilistic. Why? Because we don't know where the particle started out. Since we are ignorant of the particle's initial position, we cannot predict exactly where it will be. The hidden variable we're ignorant of is the exact position of the particle.

John Bell later proposed that de Broglie's theory should be called a theory of *beables*, as opposed to quantum theory, which is a theory of observables.[2] A beable is something that exists at all times, unlike an observable, which is a quantity evoked into existence by an experiment. In de Broglie's theory, both the particle and the wave are beables. In particular, a particle always has a position, even if quantum theory cannot predict it precisely.

Nonetheless, de Broglie's picture of a quantum world where particles and waves are both real did not catch on. In 1932, the great mathematician John von Neumann published a book in which he proved that hidden variables were impossible.[3] A few years later, a young German mathematician named Grete Hermann pointed out that von Neumann's proof had a big hole in it.[4] He had apparently committed the fallacy of assuming what he wanted to prove and had fooled himself and others by cloaking the assumption in a technical axiom. But her paper was ignored.

It took two decades for the error to be rediscovered. The American quantum physicist David Bohm wrote a textbook on quantum mechanics in the early 1950s.[5] Ruminating on the mysteries of quantum theory, he reinvented de Broglie's hidden variables theory — of which he had been ignorant. He wrote a paper describing the new quantum theory, but when he submitted it to a journal he received a referee report rejecting the paper because it disagreed with von Neumann's well-known proof of the impossibility of hidden variables. Bohm quickly found the error in the proof and wrote a paper pointing it out.[6] Since then, the de Broglie–Bohm approach to quantum mechanics — as it is now called — has been pursued by a small number of specialists; it is one of the approaches to the foundations of quantum theory still actively pursued today.

With the de Broglie–Bohm theory, we understand that hidden-variables theories are a possible option for the solution of the puzzles of quantum theory. Its study has proved useful, because many of its features have been shown to apply to any possible hidden-variables theory.

The de Broglie–Bohm theory has an ambivalent relationship to relativity theory. The statistical predictions it makes agree with quantum mechanics and can be made compatible with special relativity — and, in particular, with the relativity of simultaneity. But unlike quantum mechanics, the theory does more than make statistical predictions; it gives a detailed physical picture of what goes on in each individual experiment. The wave, which evolves in time, influences where the particle travels; in so doing, it violates the relativity of simultaneity, because the law by which the wave influences the particle's motion can be true only in one observer's frame of reference. Thus, to the extent that we take the deBroglie–Bohm hidden-variables theory as an explanation for quantum phenomena, we have to believe that there is a preferred observer, whose clocks measure a preferred notion of physical time.

This ambiguous relationship with relativity turns out to extend to any possible hidden variables theory.[7] Such a theory's statistical predictions, which agree with quantum mechanics, will agree with relativity theory. But any more detailed picture of individual events will violate the relativity principle and be interpretable only within a single observer's viewpoint.

The de Broglie–Bohm theory has one big flaw, which is that it fails to satisfy one of our criteria for a cosmological theory — the requirement that all actions be reciprocal. The wave influences where the particle goes, but the particle has no influence on the wave. Because of this, the theory is unsatisfactory as a cosmological theory. There is, however, an alternative hidden-variables theory that eliminates this problem.

◆

As a believer in Einstein's view that there must be a deeper theory behind quantum theory, I have since my student days tried to invent hidden-variables theories. Every few years, I put my other research aside and attempt to solve this crucial problem. For many years, I worked on an approach based on a hidden-variables theory written down by the Princeton mathematician Edward Nelson. These attempts worked, but they all had an element of artificiality, in that certain forces had to be exquisitely balanced to reproduce the predictions of quantum mechanics. In 2006, I wrote a paper explaining the technical reasons behind this artificiality[8] and then abandoned the approach.

One afternoon in early fall of 2010, I went to a cafe, opened a notebook to a blank page, and thought about my many failed attempts to go beyond quantum mechanics. I began by thinking about a version of quantum mechanics called the *ensemble interpretation.* This interpretation ignores the futile hope of describing what goes on in an individual experiment and instead describes an imaginary collection of all the things that *might* be going on in the experiment. Einstein put it nicely: "The attempt to conceive the quantum-theoretical description as the complete description of the individual systems leads to unnatural theoretical interpretations, which become immediately unnecessary if one accepts the interpretation that the description refers to ensembles (or collections) of systems and not to individual systems."[9]

Consider the lone electron orbiting a proton in a hydrogen atom. According to the proponents of the ensemble interpretation, the wave is associated not with the individual atom but with the imaginary collection of copies of the atom. In different members of this collection, the electrons have different positions. Thus, if you were to observe the hydrogen atom, the result would be as if you had picked out an atom at random from this imaginary collection. The wave gives the probabilities of finding the electron in all those different places.

I had liked this idea for a long time, but all of a sudden it seemed totally crazy. How could an imaginary collection of atoms influence a measurement made on one real atom? This would contradict the principle that nothing outside the universe can act on something inside the universe. So I asked myself whether I could replace that imaginary collection with a collection of real atoms. Being real, they would have to exist somewhere in the universe. Well, there are in fact a great many hydrogen atoms in the universe. Could they be the "collection" that the ensemble interpretation of quantum mechanics refers to?

Imagine that all the hydrogen atoms in the universe play a game together. In this game, each atom recognizes which other atoms are in a similar situation with a similar history. By "similar," I mean that they would be described probabilistically by the same quantum state. Two particles in the quantum world can have identical histories, and so be described by the same quantum state, but differ in the precise values of their beables, such as position. When an atom does recognize another atom as having a similar history, it copies its properties, including the exact values of its beables. There's no need for the two atoms to be close to each other for one to copy the other's properties; they just both have to exist somewhere in the universe.

This is a highly nonlocal game, but we know that any hidden-variables theory has to express the fact that quantum physics is nonlocal. Although the idea may sound crazy, it may be less crazy than having imaginary collections of atoms influence the real atoms in the world. So I decided to play the idea out and see where it went.

One of the properties that gets copied is where the electron is, relative to the proton. So the position of an electron in a particular atom will be jumping around as it copies the positions of electrons in other atoms in the universe. As a result of all this jumping, if I measure where the electron is in a particular atom, it will be as if I had picked an atom randomly from the collection of all similar atoms. So the quantum state is replaced by the collection of similar atoms. To make this work, I invented rules for the copying game that led to the probabilities for the atom to respond to measurement exactly as it would according to quantum mechanics.[10]

And I realized something that pleased me immensely: What if a system has *no* copies in the universe? Then the copy game cannot go on, and quantum mechanics will not be reproduced. This would explain why quantum mechanics does not apply to big complex systems, such as cats and you and me: We are unique. It resolves the longstanding paradoxes that arise if you apply quantum mechanics to big things like cats and observers. The strange properties of quantum systems are restricted to atomic systems because those come in a great many copies in the universe. The quantum uncertainties arise because these systems are continually copying one another's properties.

I call this the *real-ensemble interpretation* of quantum mechanics, but in my notes it's the "White Squirrel" interpretation, named for a lone albino squirrel that has been sighted in some of Toronto's parks. You can imagine that all the gray squirrels are identical enough that quantum mechanics applies to them — look to see where one is, and you might see another and another. But the white squirrel perched momentarily on a tree branch seems to have no copies and hence is not quantum mechanical. Like you or me, it may be seen as having unique properties not shared with or copied from anything else in the universe.

The game of jumping electrons violates special relativity. The jumps cross arbitrarily large distances instantaneously, hence they require a notion of simultaneous events separated by large distances. This in turn requires faster-than-light transmission of information. Nonetheless, the statistical predictions reproduce those of quantum theory and so can be made consistent with relativity. But when you look behind the scenes, there is a preferred simultaneity and hence a preferred time, just as in the de Broglie–Bohm theory.

In both hidden-variables theories I have described, the principle of sufficient reason is satisfied. There is a detailed picture of what happens in individual events which explains what quantum mechanics views as uncertain. But the price — the violation of the principles of relativity theory — is high.

◈

Could there be a hidden-variables theory compatible with the principles of relativity theory? We know that the answer is no. If there were such a theory, it would violate the free-will theorem — a theorem implying that there's no way to determine what a quantum system will do (hence no hidden-variables theory) as long as the theorem's assumptions are satisfied. One of those assumptions is the relativity of simultaneity.

The aforementioned theorem of John Bell also rules out local hidden-variable theories — local in the sense that they involve only communication at less than the speed of light.

But a hidden-variables theory *is* possible, if it violates relativity.

As long as we're just checking the predictions of quantum mechanics at the level of statistics, we don't have to ask how the correlations were actually established. It is only when we seek to describe how information is transmitted within each entangled pair that we need a notion of instantaneous communication. It's only when we seek to go beyond the statistical predictions of quantum theory to a hidden-variables theory that we come into conflict with the relativity of simultaneity.

To describe how the correlations are established, a hidden-variables theory *must embrace one observer's definition of simultaneity*. This means, in turn, that there is a preferred notion of rest. And that, in turn, implies that motion is absolute. Motion is absolutely meaningful, because you can talk absolutely about who is moving with respect to that one observer — call him Aristotle. Aristotle is at rest. Anything he sees as moving is really moving. End of story.

In other words, Einstein was wrong. Newton was wrong. Galileo was wrong. There is no relativity of motion.

This is our choice. Either quantum mechanics is the final theory and there is no penetrating its statistical veil to reach a deeper level of description, or Aristotle was right and there is a preferred version of motion and rest.

14

Time Reborn from Relativity

WE HAVE SEEN THAT the reality of time opens up new approaches to understanding how the universe chooses its laws while making possible a new resolution of the mysteries of quantum mechanics. But we still have to surmount a big obstacle, which is the formidable argument from special and general relativity in favor of the block-universe picture. This argument concludes that what is real is only the history of the universe as a timeless whole.[1]

The argument for the block universe rests on the relativity of simultaneity, which is an aspect of the theory of special relativity (see chapter 6). But if time is real, in the sense of a real present moment, there is a boundary all observers can agree on between the real present and the not yet real future. This implies a universal, physical notion of simultaneity that includes distant events and, indeed, the whole universe. This can be called a *preferred global time* ("global" here meaning that the definition of time extends throughout the universe). We have a direct confrontation between an argument that there should be a preferred global time and an argument that the principles of relativity theory preclude this. Also, as we saw in the last chapter, a preferred global time is a necessary ingredient of any hidden-variables theory, which would explain the choices made by individual quantum systems. Thus, as we also saw in the last chapter, there is a conflict between the relativity of simultaneity and the principle of sufficient reason.

The purpose of this chapter is to resolve the conflict in favor of the principle of sufficient reason. This means giving up the relativity of simultaneity and embracing its opposite: that there is a preferred global notion of time. Remarkably, this does not require overthrowing relativity theory; it turns out that a reformulation of it is enough. The heart of the resolution is a new and deeper way of understanding general relativity theory which reveals a new conception of real time.

A preferred notion of global time picks out a family of observers, spread though the universe, whose clocks measure it. This implies a preferred state of rest, reminiscent of the Aristotelian notion of rest, or the aether of 19th-century physics, both of which Einstein punctured with his invention of special relativity. For physicists before Einstein, this aether was necessary, because light waves needed a medium within which to propagate. Einstein demolished it, because his principle of the relativity of simultaneity implies that there is no aether, no state of being at rest.[2]

There is not only a contradiction here but also cause for depression. The elimination of the aether was a great triumph of focused reasoning over lazy habits of thought. It was so easy to think of the world in Aristotle's terms. Galileo and Newton established the relativity of inertial frames, which made it impossible to detect a preferred state of rest by watching how bodies move. But the idea of rest as natural still lurked silently in the minds of physicists, giving a place for the aether to roost when theorists needed a medium for light to propagate in. Only Einstein had the insight needed to demolish it altogether. And yet it seems we have reasons to go back to the idea of a preferred global notion of time. The fact that this contradicts the triumph of Einstein over the aether is a psychological barrier to taking the arguments for the reality of time seriously — or at least it was in my case.

Before discussing how theory might resolve this contradiction, let's look at what experiment has to say. A preferred global notion of time implies a preferred observer, whose clock measures that preferred time. This contradicts the relativity of inertial frames, according to which there is no experimental or observational way to distinguish an observer supposedly at rest from those moving with a constant but arbitrary velocity.

The first thing to note is that the universe is arranged in a way that does indeed pick out a preferred state of rest. We know this because when we look around with our telescopes we see the great majority of galaxies moving away from us at roughly the same speed in every direction. But this can only be true of one observer, because someone moving rapidly away from us and into space would see those galaxies ahead of her, which she is catching up with, moving slower than those behind her. Moreover, we have good evidence that the galaxies are uniformly distributed in space, at least when their positions are averaged over a sufficiently large scale — that is, the universe seems to be the same when looked at in any direction. From these facts, we can deduce that at each point in space there will be one special observer who sees the galaxies moving away from her at the same speed in every direction.[3] So the motions of the galaxies pick out a preferred observer, and hence a preferred state of rest, at each point in space.

Another way to fix a preferred family of observers is to use the cosmic microwave background. These preferred observers see the CMB coming at them at the same temperature from all directions in the sky.[4]

Happily, the two families of preferred observers coincide. The galaxies appear, on average, to be at rest in the same reference system in which the CMB comes at us at the same temperature from all directions. So the universe is organized in a way that picks out a preferred state of rest. But this fact need not contradict the principle of the relativity of motion. A theory can have a symmetry that is not respected by its solutions. To the contrary: Solutions of theories often break the theories' symmetries. The fact that there is fundamentally no preferred direction in space does not prevent the wind from blowing from the north today. Our universe represents just one solution to the equations of general relativity. That one solution can be asymmetric — that is, it can include a preferred state of rest — without contradicting the principle that the theory has a symmetry. The universe might have started off in a way that broke the symmetry.

On the other hand, we do want to ask why the universe is in a special state such that it clearly picks out a preferred family of observers. This is another question about why the initial conditions of the universe were so special. It's a question that general relativity by itself cannot answer — another clue that there's something about the universe that isn't captured by general relativity. So it's worth considering the possibility that the preferred state of rest in the universe represents something deeper. Perhaps it tells us something about a level of physics below that of general relativity.

If the existence of a preferred state of rest in the universe represents something deeper, then that state should show up in other kinds of experiments. But on scales smaller than the cosmological, the principle of the relativity of inertial frames is pretty well tested. A formidable amount of experimental evidence confirms the predictions of Einstein's theory of special relativity, much of which can be understood as testing whether or not there is a preferred state of rest in nature.[5]

So the observations give us a mixed message. On the largest scale, there is evidence for a preferred state of rest, which must be explained by something special in the initial conditions of the universe. But on every smaller scale, the evidence is that the principle of relativity rules. An ingenious solution to this conundrum has only recently been devised. General relativity, it turns out, can be reformulated in a beautiful way as a theory with a preferred notion of time. This reformulation is just another way to understand general relativity, but it reveals a physically preferred synchronization of clocks throughout the universe. Furthermore, the choice of that preferred synchronization depends on the distribution of matter and gravitational radiation throughout the universe, so it is not a throwback to Newton's absolute time. Nor can it be discovered by any local measurements, so it is completely compatible with the relativity principle for small subsystems of the universe.

The theory that enables this reversal of perspective is called *shape dynamics*.[6] Its main principle is that all that is real in physics is connected with the shapes of objects, and all real change is simply changes in those shapes. Size means nothing, fundamentally, and the fact that objects seem to us to have an intrinsic size is an illusion.

Shape dynamics was created by following a chain of thought proposed by Julian Barbour, whose timeless quantum cosmology was discussed in chapter 7. Barbour is a great advocate of the relational philosophy, and the road to shape dynamics began with his insistence on making physics as relational as possible. Many of the key steps were taken by him over the last decade, with Niall Ó Murchadha and several young collaborators, but the final pieces were put in place in the summer and fall of 2010 by a trio of young people working at Perimeter Institute: graduate students Sean Gryb and Henrique Gomes and a postdoc, Tim Koslowski.[7]

Once you know the basic ideas of relativity theory, it's easy to understand shape dynamics, because that theory is a natural next step. Let's recall some aspects of simultaneity: It's meaningful to talk about two adjacent events occurring simultaneously. We can also order them in time, and it makes sense to do so, because one event could be the cause of the other. But when we try to order events that are far from each other, we find there is no absolute ordering that all observers can agree on. For some observers, the two events may be simultaneous; for other observers, one event may appear to be in the past of the other.

Barbour tells us that size behaves the same way. If we have two adjacent objects, it makes sense to order them by size: If you can put a mouse into a box, it makes sense to say that the mouse is smaller than the box. If you have two soccer balls, it makes sense to say they have the same diameter. These comparisons make physical sense, and all observers will agree about them.

But now let's ask if the mouse here is smaller than a box in the next galaxy. Does this question still make sense? Does it have an answer all observers will agree on? The problem is that because they're far away from each other, you cannot put the mouse into the box to see whether it fits.

To answer the question, you can move the box to where the mouse is and see if the mouse fits inside. But this answers a different question, because now the box and the mouse are in the same place. How do we know there's not some physical effect expanding everything we move into our galaxy, so that a box the size of a mouse's eye becomes, en route, large enough to enclose the mouse? We might leave the box where it is and instead send a ruler to it. But how do we know that the ruler doesn't suffer the reverse effect, shrinking as it travels from mouse to faraway box?

This is one line of thought that led Barbour and friends to propose that it's not sensible to compare the sizes of objects distant from one another. What you can do is compare shapes, because shapes are not subject to the same kind of arbitrary modification. The one exception to the relativity of size is that the volume of the whole universe at each moment of time must remain unchanged. It's not easy to explain this in nontechnical language, but what it means is that if you shrink everything at one place, there must at the same time be somewhere else where you compensate by enlarging everything by the same amount, so the overall volume of the universe doesn't change. The volume can still of course change in time as the universe expands.

Although shape dynamics is radical when it comes to sizes, it's conservative when it comes to time. There's a single rate at which time flows. It's the same throughout the universe, and you aren't allowed to alter it.

General relativity is more or less the opposite. Sizes of objects are fixed and remain fixed when you move them around, so it's meaningful to compare the sizes of distant things. What general relativity is flexible about is the rate of time. It is not meaningful to ask whether a clock distant from us is running fast or slow with respect to a clock near us, because the speeding-up and slowing-down of distant clocks are among the funhouse changes that observers will disagree about. Even if you synchronize your watch with a distant clock, they can go out of sync, because there's no physical meaning to their rates staying the same.

In a word, in general relativity size is universal and time is relative, whereas in shape dynamics time is universal and size is relative. Remarkably, though, these two theories are equivalent to each other, because you can — by a clever mathematical trick that isn't necessary to go into here — trade the relativity of time for the relativity of size. So you can describe the history of the universe in two ways, in the language of general relativity or the language of shape dynamics. The physical content of the two descriptions will be the same, and any question about an observable quantity will have the same answer.

When that history is described in the language of general relativity, the definition of time is arbitrary. Time is relative and there's no meaning to what it is at distant locations. But when the history is described in the language of shape dynamics, a universal notion of time is revealed. The price you pay is that size becomes relative and it becomes meaningless to compare the sizes of objects far from one another.

Like the wave/particle picture of quantum theory, this is an example of what physicists call a duality — two descriptions of a single phenomenon, each of which is complete yet incompatible with the other. This particular duality is one of the deepest discoveries of contemporary theoretical physics. It was proposed in a different form[8] in 1995 by Juan Maldacena in the context of string theory and has since become the most influential idea in that field. As of this writing, the exact relationship between shape dynamics and Maldacena's duality is unclear, but it seems likely that there's a correspondence.[9]

Whereas there's no preferred time in general relativity, there is one in the dual theory. We can use the fact that the two theories are interchangeable to translate time in the shape dynamics world to the general relativity world. There it reveals itself as a preferred time, hidden in the equations.[10]

This global notion of time implies that at each event in space and time there is a preferred observer whose clock measures its passage.

But there is no way to pick out that special observer by any measurements made in a small region. The choice of the special global time is determined by how matter is distributed across the universe. This coincides with the fact that experiments agree with the principle of relativity on scales smaller than that of the universe. Thus, shape dynamics achieves an accord between the experimental success of the principle of relativity and the need for a global time demanded by theories of evolving laws and hidden-variable explanations of quantum phenomena.

As noted, one quantity not allowed to change when you expand and shrink scales is the overall volume of the universe at each time. This makes the overall size of the universe and its expansion meaningful, and this can be taken for a universal physical clock. Time has been rediscovered.

15

The Emergence of Space

THE MOST MYSTERIOUS aspect of the world is right in front of us. Nothing is more commonplace than space, yet when we examine it closely, nothing is more mysterious. I believe that time is real and is essential for a fundamental description of nature. But I think it likely that space will turn out to be an illusion of the sort that temperature and pressure are — a useful way to organize our impressions of things on a large scale but only a rough and emergent way to see the world as a whole.

Relativity theory merged space with time, leading to the block-universe picture, in which both space and time are understood to be subjective ways of dividing up a four-dimensional reality. The hypothesis of the reality of time frees time from the false constraints of this unification. We can develop our ideas of time with the understanding that time is very different from space. This separation of time from space liberates space, too, opening a door to a greater understanding of its nature. As we'll see in this chapter, it leads to the revolutionary insight that space, at the quantum-mechanical level, is not fundamental at all but emergent from a deeper order.

The simple fact that the world of everyday objects is organized in terms of "near" and "far" is a consequence of two basic features of reality: the existence of space and the fact that things have to be in our neighborhood to affect us (the property that physicists call locality). The world is full of things that represent either danger or opportunity, yet at any one time you aren't concerned with most of them. Why? Because they're far away from you. The tigers in countries across the ocean would eat you in a minute if they could, but you needn't worry, because they're nowhere near you. This is the gift of space; almost everything is far away from us and can be ignored for the time being.

Imagine a world containing an enormous variety of objects without the organization of space. Anything might impinge on anything else at any time. There would be no distance to keep things separated.

We are acutely aware, through our senses, of what's close to us. But not much is. It is a feature of space that not many things can occupy the spaces closest to you. This is a consequence of space's low dimensionality. Think of how many neighbors live directly next to you. Just two families, one on each side. Now, how many neighbors could be directly adjacent to you? Four — two families next door, one across the street, and one in back. If you live in an apartment building, the number of neighbors closest to you grows to six, because there are also the people underneath you and the college kids who listen to television until three in the morning just above. The number of neighbors nearest you grows proportionally to the number of dimensions — two in one dimension, four in two dimensions, six in three dimensions. The relation is simple: The number of neighbors is twice the dimension.

So if we lived in a fifty-dimensional space, we could have a hundred nearest neighbors. As it is, stuck in three dimensions, if we want to live in a building with a hundred other families, it has to be a big apartment building, and most of them will not be close neighbors. In three dimensions, we have neighbors we never meet.

This is, by the way, a problem in planning a scientific institute, where we want to maximize the chances for serendipitous interaction between people with different ideas and interests. When Perimeter Institute first opened, with seven scientists, this was no problem;

now that we have more than a hundred, it's a challenge. As theoretical physicists, we did contemplate increasing the number of dimensions in the building as we grew, but we couldn't get the architects to go for it.[1]

The fact is, we're stuck in a low-dimensional world. This, more than anything else, keeps us safe from tigers, insomniac neighbors with TV sets, and other beasts, but it's also the primary obstacle we face in attempting to increase our opportunities.

Before technology, the fact that the Earth's surface is two-dimensional kept people relatively isolated. Most people met only a few hundred others in a lifetime — those within walking distance. They did their best, arranging feasts and festivals to increase interaction (just as scientists do) with neighboring villages, and a few intrepid traders ventured abroad. But space made strangers of almost all of us.

Now we live in a world in which technology has trumped the limitations inherent in living in a low-dimensional space. Consider just the effect of cell phones. I can pick up mine and instantly be talking to almost anyone else, because 5 billion of the 7 billion people on the planet have a mobile. This technology has effectively dissolved space. From a cell-phone perspective, we live in a 2.5-billion-dimensional space, in which very nearly all our fellow humans are our nearest neighbors.

The Internet, of course, has done the same thing. The space separating us has been dissolved by a network of connections that essentially brings everyone closer. In effect, we live together in a higher-dimensional space. We're fast becoming a world in which many people may choose to live almost exclusively in that higher-dimensional space. All that's needed is a bit more virtual reality — so that, say, a cell-phone call will summon up a hologram of the person you're calling and project your hologram wherever they are.

In a high-dimensional world with unlimited potential for connection, you're faced with many more choices than in the physical world of three dimensions. Many of the challenges facing our wired world are due to this vastly enlarged sea of possibilities, and many of the social media that have proliferated are designed to exploit and manage them.

Imagine a child brought up in that high-dimensional world, where space plays no role. They will think of their world as a vast network in which a fluid and dynamic system of connections keeps everyone one step away from everyone else. Imagine, now, that someone pulls the plug. The power fails, and the citizens of the network fall into a more constrained and less stimulating world. They discover that they really live in three dimensions and that space separates most people. The number of neighbors shrinks from 5 billion to a handful, and almost everyone is suddenly very far away.

This image is a metaphor for how some physicists are now thinking about space. We (yes, I am among them) believe that space is an illusion and that the real relationships that form the world are a dynamical network a bit like the Internet or cell-phone networks. We experience the illusion of space because most of the possible connections are off, pushing everything far away.

This picture emerges from a class of approaches to quantum gravity in which space is not taken to be fundamental, but time is. These approaches posit a fundamental quantum structure — one that doesn't need space to define it. The idea is that space emerges, just as thermodynamics emerges from the physics of atoms. Such approaches are background-independent, because they do not assume the existence of a fixed-background geometry. Rather, the primitive notion is of a graph or network, defined intrinsically, without reference to space.

The first of these approaches to be developed is called *causal dynamical triangulations*, invented by Jan Ambjørn and Renate Loll and refined with their collaborators.[2] This approach was followed by *quantum graphity* (so called because it proposes that the fundamental entities in nature are graphs), invented by Fotini Markopoulou[3] and explored with her collaborators.[4] The intuitive picture I have just given, of space emerging from turning off connections in a network, fits most closely into it. A third approach — in which there is a global time, which is taken as fundamental but in which space is not emergent — was introduced by Petr Horava.[5] Certain approaches to string theory, called matrix-model approaches, also can be described this way.[6]

By taking time as fundamental, these approaches differ from older background-independent approaches positing that spacetime — all together, as in the block universe — must emerge from a more fundamental description in which neither space nor time is primitive. These include *loop quantum gravity, causal sets,* and some other approaches to string theory.

There are lessons to be learned from the successes and failures of each set of approaches. These constitute the story to be told in this chapter.

A useful metaphor arising in several approaches to quantum gravity is to imagine that space is not continuous but a lattice of discrete points (see Figure 13). Particles live on the sites of the lattice and move by hopping to the nearest neighbors. Two particles exert a force or influence on each other only if they are neighbors. If the lattice is low-dimensional, the number of particles available for interaction is small; it goes up, as in our discussion of human neighbors, with the dimensionality.

We can imagine light as photons that travel by hopping, neighbor by neighbor, along the lattice. To send a photon to a faraway particle involves many hops and hence takes time.

Now think of a world on a network that has many more connections. Things are closer to one another, in the sense that it takes fewer

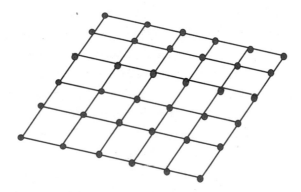

Figure 13: Space as a lattice of points. A particle can only be at one of the nodes, and motion consists of jumping from node to node.

steps to connect through the network, so it takes less time to send a signal between any two nodes of the network.

One of our principles for a new cosmology stipulates that nothing should act without being acted on. So if the network tells the particles how to move, shouldn't the network also change because of where the particles are? This would constitute an image of a physical world not so different from our interconnected human world. The world is a dynamical network of relationships; whatever lives on the network and the structure of the network itself are both subject to evolution. This is how the world is envisioned in background-independent approaches to quantum gravity.

The theory of loop quantum gravity is the oldest and best developed of the background-independent approaches to quantum gravity, so let's start the story with it. Loop quantum gravity describes space as a dynamical network of relationships. A typical quantum state of the geometry of space can be depicted as a graph — that is, a figure containing many edges, which join nodes, or vertices (see Figure 14). The edges (which indicate some kind of primitive relationships among nodes) all have labels on them specifying the relations between the nodes they connect. These labels can be taken as ordinary integers, one integer labeling each edge. (The nodes have labels, too, but these have a more complicated description, and I won't trouble the reader with that here.)

Recall that in quantum physics, the energy of an atom is quantized and only certain states with certain discrete energies have definite energy values. According to loop quantum gravity, the volumes of regions of space are also quantized; they can have only certain discrete values of the volume. The areas of surfaces are also quantized.[7] Loop quantum gravity gives precise predictions for the spectra of volume and area. These have potentially observable consequences — for example, they imply precise predictions for the spectra of radiation that might be observed emanating from small black holes.[8]

Consider a piece of steel — say, a sewing needle. It looks smooth enough, but we know that it's made of atoms in a regular arrangement.

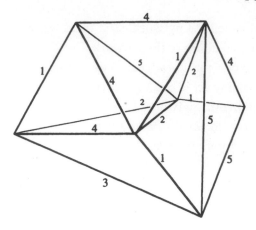

Figure 14: A typical quantum state of the geometry of space, depicted as a graph.

If we peer down to the scale of the atoms themselves, the smoothness of the metal gives way to a picture made up of discrete units — atoms — connected to one another in a regular way. Space also appears "smooth," or continuous, but if loop quantum gravity is right, then space, too, is made of discrete units, which can be thought of as "atoms" of space. If we could observe at the Planck scale, we would see the smoothness of space transformed into this picture.

In general relativity, as we have seen, the geometry of space turns out to be dynamical. It evolves in time, in response to matter moving or gravitational waves propagating. But if geometry is really quantum at the Planck scale, the changes in the geometry of space must come from changes occurring at that scale. There must, for example, be oscillations in the quantum geometry of space corresponding to the passage of a gravitational wave. A triumph of loop quantum gravity is that the dynamics of spacetime, which is given in general relativity by Einstein's equations, can indeed be coded in simple rules for how the graphs evolve in time[9]. These are illustrated in Figure 15.

This coding of the Einstein equations into rules for graphs to change works both ways. You can start with Einstein's theory and follow a procedure for turning a classical theory into a quantum theory. This procedure has been developed and tested on many different theories. Ap-

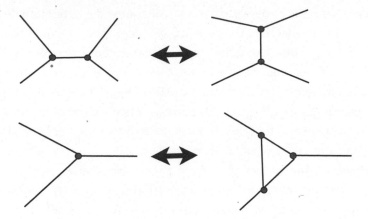

Figure 15: Rules for how graphs evolve in time in loop quantum gravity. Each move can act on a small part of the graph, as shown.

plying it to general relativity is a technically challenging exercise, but when carried out correctly it leads to the picture we have been describing here, with precise rules for the graphs to change in time. In this way, we call loop quantum gravity the "quantization" of general relativity.[10]

Alternatively, you can start with the quantum rules for the graphs to change and ask if the rules for classical general relativity can be derived as an approximation to them. This is analogous to deriving the equations that describe the flowing of water from the fundamental laws followed by the atoms making up the water. This exercise is called deriving the classical theory from the classical limit of the quantum theory. It's a challenging one, but recently there are positive results in loop quantum gravity.[11] These employ a spacetime approach to quantum spacetime called a spin-foam model, in which the network underlying the geometry of space is taken to be a part of a bigger network encompassing space and time. Hence the spin foam gives a quantum version of the block-universe picture, in which space and time are unified in a single structure. What is especially impressive is that there are several independent results showing general relativity emerging from spin-foam models.

It's even easy to add matter to the quantum geometric picture. The story is the same as the lattice model, only now the lattice can change. We can put particles on the nodes, or vertices. They move by jumping from node to node, along the edges, just as in the lattice model. If you look from far enough away, you don't see nodes or graphs, you just see the smooth geometry they approximate. The particle then looks as if it is traveling through space. So perhaps when we throw a ball, what's really happening is that the atoms in the ball are hopping from atom of space to atom of space to atom of space.

However, the results that show general relativity emerging from loop quantum gravity, as important as they are, come with some limitations. In some cases, the description is limited to a small region of spacetime surrounded by a boundary. The presence of the boundary tells us that loop quantum gravity is best conceived as a description of a small region of spacetime and hence fits into the Newtonian paradigm.

There are also results in string theory suggesting that spacetime can emerge in a bounded region — at least when the cosmological constant takes on a negative value. These arise in the context of the duality between general relativity and the scale-invariant theory conjectured by Juan Maldacena that I mentioned in chapter 14. If his conjecture is correct — and many results support it — then classical spacetime may emerge in the interior of a region whose boundary has a fixed classical geometry.

Thus both loop quantum gravity and string theory suggest that quantum gravity can be understood as describing regions of spacetime with boundaries, hence fitting within the Newtonian paradigm. Their strongest results are achieved in the context of physics in a box, without addressing the issue of whether or not the description can be scaled up to a theory of the whole, closed universe.

Another assumption of the loop-quantum-gravity results on the emergence of spacetime is that the graphs describing the quantum geometry of space are limited to those that already look like a discrete picture of a low-dimensional space.[12] In these cases, the locality of space is captured by having each vertex, or node, in the graph connected to only a small number of other vertices. Just as in a suburban neighborhood, each node has only a few nearest neighbors. To travel between two widely separated nodes, a particle has to make many hops. It takes time for a particle, or a quantum carrying information, to go a long way. Hence there emerges a description of the world with a finite speed of light. However, there are many states of quantum geometry in which there is no nice version of locality. There are graphs in which every node is connected to every other node in just a few steps. So far, loop-quantum-gravity methods do not illuminate how these quantum geometries evolve.

Consider an example in two spatial dimensions — just a large region of a plane, as pictured in Figure 13. This plane can have a quantum-geometric description, described in the figure in terms of a graph. Now, consider two nodes that are many steps away from each other in the graph; we'll call these Ted and Mary. We can make a new graph from the old one by adding another edge, which connects Ted directly to Mary (see Figure 16). This depicts a quantum geometry in which Mary and Ted are neighbors. It's as if they had both just bought cell phones; the space separating them has dissolved.

If geometry really is quantum, then there are perhaps 10^{180} nodes within our observable universe — that is, one node per Planck length cubed. If each is connected only to a few nearby neighbors, then the quantum geometry can look, at large scales, just like a classical geom-

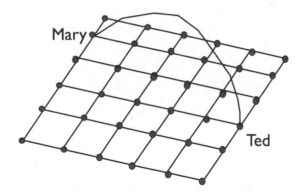

Figure 16: The addition of a nonlocal link disorders locality, bringing two faraway points close to each other.

etry. The locality of space then emerges from the particular design of the quantum geometry reproducing it. In this case, there are roughly the same number of edges as nodes, because each node is connected to just a few neighbors. But by adding just one more edge to the vast number of edges making up the quantum geometry, we violate the locality of space in a drastic way, making it possible for widely separated nodes like Ted and Mary to communicate essentially instantaneously. We call this *disordering locality,* and we call the edge that was added a *nonlocal link.*[13]

It's remarkably easy to disorder locality by adding just one nonlocal link. A single nonlocal link would be one out of 10^{180} edges within the observable universe, but there are 10^{360} possible places to insert it. If you add it randomly to a graph with 10^{180} nodes, you're much more likely to be adding a nonlocal link than a local link, because the number of ways to add a nonlocal one is so much greater than the number of ways to add it locally. The node at one end can be connected to a small number of other nodes, if you want to make a local connection. But if you don't care about locality, the other end could hitch up to any node in the universe. We see, again, what an enormous constraint locality is.

You might wonder how many nonlocal links could be added to the quantum geometry of space before we take notice in the macro world. Because ordinary particles have quantum wavelengths many orders of magnitude larger than the Planck scale, the probability of a photon of visible light finding itself at the end of a nonlocal link so that it can hop directly from Ted to Mary is very small. Rough estimates suggest that as many as 10^{100} such nonlocal links can be added before experiments will easily detect a faster-than-light communication. This is a huge number (but nowhere near as large as 10^{180}). Still, nodes connected nonlocally to somewhere across the universe would be fairly common; there would be, on average, more than one per cubic nanometer of space.

Once we allow nonlocal links, there are a vast number of ways that locality can be disordered. You could also have a few nodes that connected to many other nodes. These very social nodes would act as gossips do in a social network, channeling a lot of information throughout the universe, acting as shortcuts.

Might the universe be full of such nonlocal connections? How might their presence be detected?

An obvious idea is that entanglement and other manifestations of nonlocality in quantum theory are instances of disordered locality. Perhaps the fundamental level of description — in which there is no space, just a network of interactions, with everything potentially connected to everything else — is the hidden-variables theory whose existence I argued for in chapter 14. If so, then quantum theory and space would emerge together.[14]

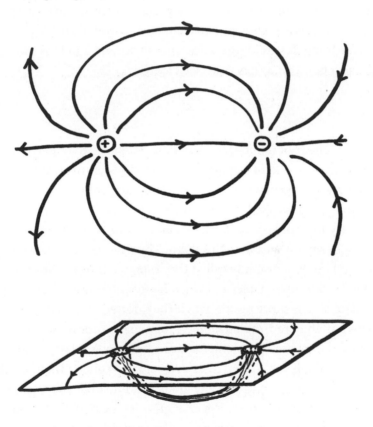

Figure 17: A long-distance link as a wormhole trapping a line of electric flux. Around one mouth of the wormhole is an electric field appearing to originate at a point, which looks like a charged particle.

Another (and only mildly crazy) hypothesis is that the nonlocal links explain the mysterious dark energy that is causing an increase in the rate at which our universe is expanding.[15] An even bolder and less likely hypothesis is that they might explain the dark matter.[16] And a still wilder hypothesis is that charged particles are nothing but the ends of nonlocal links.[17] This is reminiscent of an old idea of John Wheeler's that charged particles might well be the mouths of wormholes in space, since wormholes are small (hypothetical) tunnels that connect widely separated locations in space. The field lines of an electric field end on charged particles, but they also appear to end at wormholes, where they (presumably) jump through the tunnel and come out the other end. One end would act like a particle with positive charge, the other like a particle with negative charge.[18] A nonlocal link could do the same thing. It would trap a line of electric field and look like a particle and a faraway antiparticle (see Figure 17).

◆

A small number of nonlocal connections might be tolerated and even advantageous if one of the aforementioned ideas turns out to work. But if there are too many of them, you run into problems getting space to emerge. This is called the *inverse problem.*

It's easy to approximate a particular smooth two-dimensional surface — let's say the surface of a sphere — by a network of triangles (see Figure 18). Such a graph is called the triangulation of a surface. This is what Buckminster Fuller did when he invented the geodesic dome, and there was a short period when they dotted the landscape, until people remembered the advantages of square rooms. But now let's consider the inverse problem. Suppose I give you a large number of triangles and tell you to construct a structure by gluing them together, edge to edge. I don't give you any guidance; I just tell you to assemble the surface randomly from the triangles. It's extremely unlikely that you're going to produce a sphere. You're probably going to get a crazy shape like those illustrated in Figure 19 — a shape with spikes on it or some other complicated mess.

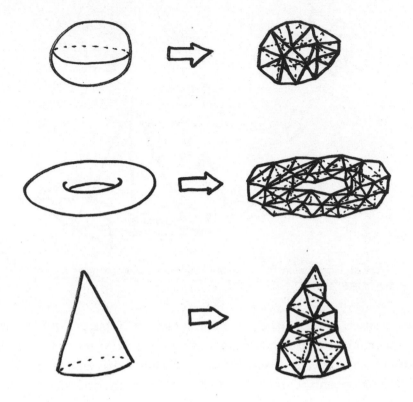

Figure 18: Triangulating smooth two-dimensional surfaces.

The problem is that there are many more ways to put triangles together to make crazy shapes than there are to make a nice two-dimensional spherical surface. In all these rogue shapes, the atomic structure sticks out, because there's a lot of complexity at the scale of the individual triangles. So nothing like a nice space emerges.

The results showing general relativity emerging from loop quantum gravity evade the inverse problem, because they're based on a particular choice of graphs which can be constructed by triangulating space. These results are, within their context, impressive, but they don't tell us how to describe the evolution of more general graphs that would have lots of nonlocal connections.

This, again, emphasizes how constraining and special is the local-

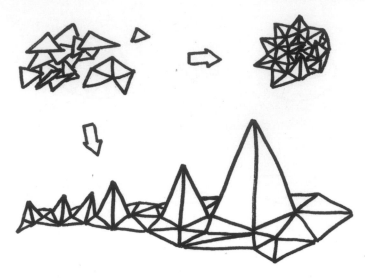

Figure 19: Crazy geometries made by randomly gluing triangles together at their edges.

ity of space. And it teaches an important lesson. If space emerges from a quantum structure, then there must be some principle or force that drives the "atoms" of space to assemble in a way that limits the possible arrangements to those that "look like" space. In particular, the fact that each space atom has only a few other space atoms in its neighborhood must be forced—because that doesn't happen in a random assembly of atoms of space.

I have been talking about quantum general relativity as constructed via loop quantum gravity, but the issue of the inverse problem afflicts other approaches to quantum gravity involving the idea of space or spacetime having what amounts to an atomic structure. These include the approaches called causal set theory, matrix models of string theory, and dynamical triangulations. Each has attractive features that have motivated people to study it, and each has run into the inverse problem.

The main question these approaches pose is why the real world looks like three-dimensional space rather than a highly interconnected network.

To appreciate the difficulty, imagine that we live in the network of cell-phone users. Space is nonexistent, and the only notion of distance, or of who is a neighbor and who is not, is defined by who calls whom. If you talk with someone at least once a day, we'll consider the two of you to be nearest neighbors. The less often you call someone, the farther away from that individual you will be. Now, notice how different and more flexible this notion of distance is than distance in space. In space, as we saw, everyone has the same number of potential nearest neighbors; in three-dimensional space, unlike in a cell-phone network, no one can have more than six.

In a cell-phone network, you are also free to be as near or as far as you like from any other user in the network. If I know how far you are from, say, 50,000 other users, that tells me nothing about how far you might be from the 50,001st. The next user added might be a stranger or might be your mother. But in space, proximities are rigid. Once you tell me who your nearest neighbors are, I know where you live. I can tell how far away you are from everybody else.

Consequently, it takes much more information to specify how a network is connected up than it does to specify how objects are arranged in a two- or three-dimensional space. To specify how connected the 5 billion cell-phone users are, I need to give a separate bit of information for every potential pair of users. That's roughly the square of 5 billion, which is written as 2.5×10^{19}. But to specify where every user is on the surface of the Earth, I need only two numbers for each: their longitude and latitude — that is, a paltry 12 billion numbers. So if space arises from turning off connections in a network, then there are a huge number of potential connections that must be shut down.

How are these connections to be turned off?

The quantum-graphity approach to quantum gravity addresses this issue by supposing that creating and maintaining connections in a network requires energy. Then it takes much less energy to form a two- or three-dimensional lattice, like Figure 13, than it does to form higher-dimensional lattices. This suggests a simple picture of the very early universe: At the beginning, it was very hot, so there was enough energy to turn most connections on. The early universe was therefore a world in which everything was connected to everything else, in at most a few steps. As the universe cooled, the connections began to fail, until only the few needed to make a three-dimensional lattice remained. This is a scenario for the emergence of space (some of my colleagues talk about the Big Freeze rather than the Big Bang). The process has also been called *geometrogenesis*.[19]

Geometrogenesis can explain some puzzling features of the initial conditions of the universe, such as why the CMB radiation comes at us from all directions with the same temperature and the same spectrum of fluctuations: It is because initially the universe was a highly interconnected system. Geometrogenesis thus provides an alternative to the hypothesis that the universe underwent a tremendous inflation early in its life.

Of course, the devil is in the details, and exactly how and why the Big Freeze should result in a three-dimensional structure that looks regular, like the two-dimensional lattice shown in Figure 13, rather than a more chaotic structure, is a question that is the subject of current research.[20]

◆

Solving the inverse problem turns out to teach us two important lessons about the nature of time.

The first is that space is more likely to emerge in models of quantum universes that assume the existence of a global time variable. This is illustrated by the dynamical-triangulation models.

A triangulation is, as noted, a surface built from joining lots of triangles together, as in a geodesic dome (see Figure 18). A three-dimensional curved space can be constructed in an analogous way, by joining tetrahedrons, which are the three-dimensional analogues of triangles. A dynamical-triangulation model uses these tetrahedrons as the atoms of space. A quantum geometry is described not by a graph but by an arrangement of tetrahedrons glued face to face.[21] Such a configuration of space evolves in time, through a set of rules, to build a discrete triangulated version of a four-dimensional spacetime (see Figure 20).

There are two kinds of dynamical-triangulation approaches: those

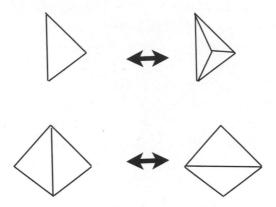

Figure 20: Evolutionary rules for triangulations of surfaces.

in which spacetime is atomized and meant to be emergent, as in the block-universe picture, and those in which a universal notion of time is assumed and only space is asked to emerge. Otherwise the constructions are very similar. The result is that a coherent spacetime emerges only in the models in which time is assumed to be real. The other models — the ones without a global time — fall victim to the inverse problem: that is, they are overwhelmed by an infection of rogue geometries that look nothing like space (see Figure 19).

The models that resolve the inverse problem are the ones known as causal dynamical triangulations, invented by Ambjørn and Loll. These emergent spacetimes are partly realistic, in that they have three dimensions of space and one of time; some of them are shown in Figure 21. They are the first examples of quantum universes that look, on large scales, like solutions of Einstein's theory of general relativity. They even demonstrate that the volume of space grows in time in the way the Einstein equations require.

There remain some issues to be resolved — for instance, whether these emergent spacetimes resemble solutions of general relativity in sufficient detail to reproduce phenomena such as gravitational waves and black holes. Another challenge is to understand the fate of the global notion of time built into the models. An old question is whether

Figure 21: A typical spacetime geometry emerging from a causal-dynamical-triangulation model.[22]

the presence of the global time violates the many-fingered-time symmetry of general relativity (see chapter 6). A newer way to ask this question is whether general relativity is — or can, by some adjustment of the model — be recovered in the form of shape dynamics, which, as we saw in chapter 14, is a theory with a global time equivalent to general relativity.

The second lesson is that if space is emergent, then there cannot be a relativity of simultaneity at the deepest level, because everything is connected to everything else. Since you can send a signal between any two nodes in one or a few steps, there is no problem synchronizing clocks. Hence, at that level, time must be global.

This lesson is illustrated by the results from quantum-graphity models. Here the setting is a graph with a large number of nodes, any two of which are either connected or not. The quantum geometries then include any graph that can be drawn connecting all the nodes. A dynamical law turns connections on or off. Several models have been studied in which different laws are assumed to turn the edges on and off. These models appear to have two phases, analogous to two phases of water. There's a high-temperature phase, in which almost all the edges are on, so every node is closely connected to every other node in one or a very few steps. There's no locality, because information can hop easily and quickly between any two nodes. In this phase of the model, there's nothing like space. But if you cool the model, there appears to be a phase transition to a frozen phase, in which almost all the edges turn off. As in a low-dimensional space, each node has just a few nearest neighbors, and there are many hops between most pairs of nodes.

You can also put matter into a quantum-graphity model. Particles live on the nodes and can hop from one node to another only when the edge connecting them is on. A dynamics can be hypothesized that captures the principle of reciprocal action also realized in general relativity — the principle that geometry tells matter where it can move and matter tells geometry how it can evolve. These models show some features of the emergence of space and also have gravitational-type phenomena, such as analogues of quantum black holes, where particles can be trapped for long periods of time. These black-hole regions are not permanent; they slowly evaporate in a way that reminds us of Stephen Hawking's process of black-hole evaporation.

There's much work yet to be done on these models before we can conclude that they may be realistic — but already, simply as toy models, they have been of enormous heuristic benefit. They show that if everything is potentially connected to everything else, then there must be a global time. The relativity of simultaneity in special relativity is a consequence of locality. Determining whether distant events are simultaneous is impossible because the speed of light imposes an upper limit on the transmission of signals. In special relativity, you can determine simultaneity only when two events are at the same place. But in a quantum universe where every particle is potentially one step away from every other particle, everything is essentially "at the same place." In such a model, there's no problem synchronizing clocks, so there is a universal time.

When space emerges in such a model, so does locality. So, also, does the existence of a speed limit for transmission of signals. (This has been shown in some detail in quantum-graphity models.[23]) As long as you look only at phenomena in the emergent spacetime and don't probe down to the scale of the spacetime atoms, special relativity will seem approximately true — which reinforces the main lesson the models and theories described in this chapter teach us: *Space may be an illusion, but time must be real.*

The development of our understanding of quantum gravity continues. There is much value in all the approaches discussed here. They each teach us something important about potential quantum-gravitational phenomena that might be observed in nature; they also teach us about the consequences of different hypotheses, the challenges each faces, and possible strategies to overcome them. The more successful approaches either fit into the Newtonian paradigm and teach us about quantum spacetimes in a box or, if they rise to the cosmological challenge, point us toward the reality of time.

16

The Life and Death
of the Universe

W E NOW TURN TO THE most important and puzzling
question we can ask about the universe: Why is it hospitable to life? We'll see that a big part of the answer is that
time is real.

If time is truly real, then there should be features of the universe
that are explicable only if we assume that time is fundamental. These
features should appear mysterious and accidental on the contrary assumption — that time is emergent. There are indeed such features;
they are captured by the observation that our universe has a history of
evolving from the simple to the complex. This gives time a strong directionality — we speak of the universe having an arrow of time. Directionality is highly unlikely in a world in which time is inessential and
emergent.

Look around. Either with the naked eye or through the most powerful telescope, we see a universe that is highly structured and complex.

Complexity is improbable. It requires explanation. Nothing can
jump immediately from a simple to a very complex organization.

Great complexity requires a series of small steps. These occur in a sequence, which implies a strong ordering of events in time.

All scientific explanations of complexity require a history, during which the levels of complexity rachet up slowly and incrementally. This is climbing Richard Dawkins's Mount Improbable.[1] So the universe must have a history, which played out in time. There's a causal order required to explain how the universe came to its present state.

According to 19th-century physicists and some of our contemporary speculative cosmologists who embrace the timeless picture, the complexity we see around us is accidental and necessarily temporary. In their view, the universe is fated to end in a state of equilibrium. This is a state, called the *heat death of the universe,* in which matter and energy are uniformly distributed throughout the universe and nothing ever happens except rare random fluctuations.[2] Most of the time these random fluctuations dissipate as soon as they appear, constructing nothing. But, as I will explain in this and the next chapters, the principles set down in chapter 10 for a new cosmological theory will help us understand why a universe with increasing complexity is natural and necessary.

So two different roads lie before us, leading to starkly different versions of the future of the universe. In the first picture, there is no future, because there is no time. Time is an illusion that is at best a measure of change, an illusion that will end when change ceases.

In the time-bound picture I propose, the universe is a process for breeding novel phenomena and states of organization, which will forever renew itself as it evolves to states of ever higher complexity and organization.

The observational record tells us unambiguously that the universe is getting more interesting as time goes on. Early on it was filled with a plasma in equilibrium; from that simplest of beginnings, it has evolved enormous complexity over a wide range of scales, from clusters of galaxies to biological molecules.[3]

The persistence and growth of all this structure and complexity is perplexing, because it rules out the simplest explanation for the structure we see — which is that it's an accidental arrangement. An accident wouldn't result in structures that persisted for billions of years whose complexity continually increased over time. As I will shortly explain, if the complexity we see around us were accidental, it would almost certainly be decreasing rather than increasing over time.

The prediction that the universe will end in a heat death is yet another step in the expulsion of time from physics and cosmology, and akin to the ancient idea that the natural state of the universe is one without change. The oldest impulse in cosmological thinking is that the world's natural state is an equilibrium — that is, a state in which, since everything is in its natural place, there is no impulse toward organization. This was the essence of Aristotelian cosmology, which, as I described in chapter 2, was based on a physics within which every essence has a natural motion: For example, earth seeks the center, whereas air's natural motion is upward.

The only reason there is still change in the earthly realm, according to Aristotle, is that there are other causes of motion, classed as imposed motions, which can move something out of its natural state. Humans and animals are sources of imposed motions, but there are others. Hot water admits air into it, and so partly takes on the natural upward motion of air and rises until it cools, at which point it expels the air and falls as rain. The ultimate source of this imposed motion is heating from the sun, which is a part of the heavenly realm. One way or another, the source of all imposed motion is the sun. Were the earthly sphere disconnected from the heavens and left to itself, everything would come to equilibrium, at rest in its natural place, and change would cease.

Modern physics has its own notion of equilibrium, which is characterized by the laws of thermodynamics. These apply to physics in a box. The context for the laws of thermodynamics is an isolated system, which exchanges neither energy nor material with its surroundings.

We have to be careful, though, not to confuse notions of equilibrium in Aristotle or Newton from the modern notion of thermodynamic equilibrium. Equilibrium in Aristotle and Newton arises from balances of forces. A bridge stands up because the forces on each girder and rivet are balanced. The notion of equilibrium in modern thermodynamics is completely different. It applies to systems with a very large number of particles and refers in an essential way to notions of probabilities.

Before we talk about the heat death of the universe, we'd better be sure to understand our terms. This means, most of all, understanding the meaning of *entropy* and the *second law of thermodynamics.*

◆

The key to understanding modern thermodynamics is that it involves two levels of description. There's the microscopic level, which is a precise description of the positions and motions of all the atoms in any particular system. This is called the *microstate.* Then there's the macroscopic level, or *macrostate* of the system, which is a broad-brush approximate description in terms of a few variables, such as the temperature and pressure of a gas. Studying a system's thermodynamics involves assessing the relationship between these two levels of description.

A simple example is a standard brick building. The macrostate in this case is the architectural drawing; the microstate is exactly where each brick goes. The architect needs to specify only that brick walls of such-and-such dimensions are built, with openings for windows and doors. He doesn't need to say which bricks go where. Most bricks are identical, so it doesn't affect the structure if two identical bricks are swapped. Thus there are a huge number of different microstates that give the same macrostate.

Let's contrast this with a building by Frank Gehry, like the Guggenheim Museum Bilbao, whose outer surface is made up of individually crafted metal sheets. To make the curved surfaces of Gehry's designs, each sheet must be different, and it matters where each one goes. The building will take the form intended by the architect only if each and every metal sheet goes in its precise place. In this case, the architectural drawing again specifies the macrostate, and where each sheet goes is the microstate. But unlike the traditional brick building, there is no freedom to tamper with the microstate. There is only one microstate that gives the intended macrostate.

The concept of how many microstates could give the same macrostate thus gives us a way to explain how Gehry's buildings are so revolutionary. This concept has a name, which is *entropy*. The entropy of a building is a measure of the number of different ways to put the parts together to realize the drawing of the architect. A standard brick building has a very high entropy. A building by Frank Gehry can have an entropy of zero, corresponding to its unique microstate.[4]

We can see from this example that *entropy is inverse to information*. It takes a lot more information to specify the design for a Gehry building because you need to tell exactly how to fabricate each piece and exactly where every piece goes. It takes much less information to specify the design of a normal brick building because all you need to know are the dimensions of its walls.

Let's see how this method works in a more typical example from physics. Consider a container full of gas consisting of a very large number of molecules. The fundamental description is microscopic: It tells where each molecule is and how it's moving. This is a huge amount of information. Then there is a macroscopic description, in which the gas is described in terms of its density, temperature, and pressure.

Specifying the density and temperature requires much less information than is needed to say where each atom is. Consequently, there's an easy way to translate from the microscopic description to the macroscopic, but not vice versa. If you know where each molecule is, you know the density and the temperature, which is the average energy of motion. But going the other way is impossible, because there are a great many different ways the individual atoms can be arranged microscopically that will result in the same density and temperature.

To translate from the microstate to the macrostate, it is useful to count how many microstates are consistent with a given macrostate. As in the examples of the buildings, this number is given by the *entropy* of the macroscopic configuration. Note that entropy defined this way is a property only of the macroscopic description. Entropy is hence an emergent property; it makes no sense to attribute an entropy to the precise microstate of a system.

The next step is to relate the entropy to probability. You do this by making an assumption, which is that all microstates are equally probable. This is a physical postulate, justified by the fact that the atoms in a gas are in chaotic motion, which tends to shuffle and hence randomize their motions. The more ways there are to make a macrostate from microstates — that is, the higher the entropy of a macrostate — the more likely it is to be realized. The most probable macrostate, given that the microstate is random, is called the state of equilibrium. Equilibrium is also the state with the highest entropy.

Take a cat apart into its constituent atoms and mix those atoms randomly in with the air in a room. There are many more microstates in which the cat's atoms are randomly mixed in the air then there are microstates where the cat is reassembled and sitting on the couch, licking its fur and purring. The cat is a highly improbable way for the atoms to be arranged, hence it has low entropy and high information, compared to a random mixing of the same atoms into the air.

Atoms in a gas move about chaotically, colliding often. When they collide, they send one another off, moving in more-or-less random directions. So time tends to shuffle the microstate. If the microstate doesn't start out random, it will be random pretty soon. This suggests that if we start from a state other than the equilibrium state, with a low entropy, the most likely thing to happen as time goes on is that the microstate becomes more random, increasing the entropy. This is the statement of *the second law of thermodynamics.*

To see how this works, we next consider a simple experiment. We need a deck of cards and a card shuffler. Let's assume that when the experiment begins, the cards are in order. After that, all that happens is that once every second the cards get shuffled by the shuffler. The experiment is to observe what happens to the order of the cards as they are repeatedly shuffled.

The cards start out ordered, but each shuffle makes the order more and more random. The entropy tends to go up. After enough shuffles, it's impossible to tell the order apart from a purely random order; consequently any memory of the initial ordering has been essentially lost.

This tendency for order to dissipate into disorder is captured by the second law of thermodynamics. In this context, the law says that shuffling a deck of cards will tend to destroy any special ordering the cards may have had initially, replacing it with random ordering.

Entropy doesn't always go up. Every once in a while, a shuffle will take the entropy down — for example, by returning the cards to the original order. It's just much more probable for a shuffle of an ordered deck to increase the entropy than decrease it. The more cards there are in a deck, the less likely it will be that a shuffle produces a complete reordering. Hence, the longer will be the intervals between shuffles that completely order the deck. Nonetheless, as long as there is a finite number of cards in the deck, there is a time by which the shuffles, taking place at one per second, are likely to have produced a complete reordering. This is called the *Poincaré recurrence time.* If you watch a system over much shorter times, you will likely only see entropy go up. But watch the system for longer than the Poincaré recurrence time and you will likely see entropy go down as well.

The story of the role of randomness in the ordering of the cards can be transferred to the gas. Ordered configurations of the atoms in the gas exist, such as configurations in which all the atoms are on one side of the box and are all moving in the same direction. These configurations are analogous to those in which all the cards are ordered. But although these ordered configurations of the atoms exist, they are much rarer than configurations in which the atoms are randomly placed throughout the box and moving in random directions.

If we start with the atoms all in one corner of the box, all moving the same way, we will see that as they move and scatter off one another, they spread across the box, filling it entirely. After some time, the positions of the atoms will be completely shuffled, so that the density of atoms in the box becomes uniform.

At roughly the same rate, the directions the atoms move in and their energies will be randomized as they collide. Eventually, most of the atoms will have close to the average energy, which is the temperature.

No matter how ordered and unusual is the configuration you start with, after a while the density and temperature of the atoms in the box will be uniform and randomized. This is the state of equilibrium. Once the gas reaches equilibrium, most likely it will stay there.

The second law of thermodynamics, in this context, says that over short times the most likely change in entropy is a positive number, or at least zero. If you start with a configuration out of equilibrium, you are starting with a configuration of lower probability and thus lower entropy. The most probable thing to happen is that the configuration is further randomized by the collisions of the atoms, hence increasing the probability of the configuration. So the entropy goes up. If you start in equilibrium, where the entropy is maximized because the configuration is already randomized, the most likely thing to happen is for the configuration to stay randomized. But if you watch the atoms over a very long period, there will, as noted, be improbable fluctuations that lead the gas to a more ordered state. The most probable of these fluctuations are subtle: just a bit more density at one place and a bit less elsewhere. Much less probable will be fluctuations that take all the atoms back to one corner of the box. But given enough time, these will occur. As long as the number of atoms is finite, there will be fluctuations leading to any configuration, no matter how rare.

But you don't have to wait to see physical effects of such fluctuations. Einstein famously used a study of the fluctuations of the molecules in a liquid to demonstrate the existence of atoms. He hypothesized that a liquid, such as water, is made of molecules in random motion, and he pondered the effect of these motions on a tiny particle, like a pollen grain, suspended in the water. The water molecules are too small too see, but their influence can be seen in the motion of the grain, which is just big enough to be seen in a microscope. The grain is knocked around by collisions with the molecules, causing it to go into a kind of random dance.

By measuring how energetically the pollen grain dances, you can deduce how many molecules impact it per second and with what force. In one of his 1905 papers, Einstein made testable predictions, later borne out, about the properties of atoms, including the number of atoms in a gram of water.[5] From this and many similar experiments, we know that such fluctuations are real and a part of the story of thermodynamics.

The fluctuations resolve a major paradox that bedeviled early studies of thermodynamics. Originally, the laws of thermodynamics were introduced without the notion of atoms or probability. Gases and liquids were treated as continuous substances, and entropy and temperature were defined without the notion of probability, as if they had a fundamental meaning. In this original formulation, the second law said simply that in any process entropy either went up or remained the same. Another law said that when entropy is maximized, a system has the same temperature everywhere.

In the middle of the 19th century, James Clerk Maxwell and Ludwig Boltzmann developed the hypothesis that matter was made of atoms moving randomly, and they attempted to derive the laws of thermodynamics from applying statistics to the motions of large numbers of atoms. For example, they proposed that temperature was just the average energy of random atomic motion. They introduced entropy and the second law much as I have here.

But most physicists back then didn't believe in atoms. Consequently, they rejected these efforts to ground the laws of thermodynamics in atomic motion and invented powerful arguments to show that the laws of thermodynamics couldn't be derived from that motion. One such argument went as follows: The laws of motion that atoms (if they exist) must obey are reversible in time (as I discussed in chapter 5). If you take a movie of a bunch of atoms moving according to Newton's laws and run the movie backward, you also have a possible history consistent with Newton's laws. But the second law of thermodynamics is not reversible, because it says that entropy always increases or stays the same but never decreases. It is impossible, the skeptics argued, that a law that is not reversible in time could be derived from laws that are—i.e., those governing the motions of the putative atoms.

The right answer to this was given by Paul and Tatiana Ehrenfest, a young couple who were protégés of Boltzmann and later became friends of Einstein.[6] They showed that the second law as formulated in pre-atomic physics was wrong. Entropy does indeed decrease sometimes, it is just not probable that it will. If you wait long enough, fluctuations will occasionally reduce a system's entropy. So fluctuations are a necessary part of the story of how thermodynamics was reconciled with the existence of atoms obeying fundamental time-reversible laws.

Nonetheless, even the correct picture seems devoid of hope for the future, because any isolated system will, on these principles, eventually come to equilibrium — after which there is no meaningful cumulative change, no growth of structure or complexity, but only an infinite equilibrium in which nothing happens but random fluctuations.

A universe in equilibrium cannot be complex, because the random processes that bring it to equilibrium destroy organization. But this does not mean that complexity itself can be measured by the absence of entropy. To fully characterize complexity we need notions beyond the thermodynamics of systems in equilibrium; these are the subject of the next chapter.

◆

When we view cosmology from the perspective of thermodynamics, the question of why the universe is interesting becomes even more puzzling. From the point of view of the Newtonian paradigm, the universe is governed by a solution to the equations of some law. That law may be approximated by some combination of general relativity and the Standard Model of Particle Physics, but the details don't matter. The solution that governs the universe is picked out from an infinite set of possible solutions and can be specified by picking initial conditions at or near the time of the Big Bang.

What we learn from thermodynamics is that almost every solution to the laws of physics describes a universe in equilibrium, because the definition of equilibrium is that it is composed of the most probable configurations. Another implication of equilibrium is that a typical solution to the laws is time-symmetric — in that local fluctuations to a more ordered state are just as probable as fluctuations to a less ordered state. Running the movie backward results in a history equally probable and on average equally time-symmetric. We can say that there is no overall global arrow of time.

Our universe looks nothing at all like these typical solutions to the laws. Even now, more than 13 billion years after the Big Bang, our universe is not in equilibrium. And the solution that describes our universe is time-asymmetric. These properties are extraordinarily unlikely, were the solution that describes our universe to be picked randomly.

The question of why the universe is interesting and appears to be getting more so is akin to the question of why the second law of thermodynamics has yet to act to randomize the universe into thermal equilibrium, in spite of billions of years of apparent opportunity to do just that.

◆

The simplest sign that our universe is not in thermal equilibrium is that there is an arrow of time. The flow of time is marked by a strong asymmetry: We feel and observe ourselves moving from the past into the future.

Countless phenomena attest to the directionality of time. Many things are irreversible (a car accident, a badly chosen phrase spoken to an insecure friend, a spilt glass of milk). Hot cups of coffee cool down rather than the reverse; sugar mixes into them, not out; and dropped cups shatter into pieces that, left to themselves, will never reassemble. We all age in the same direction; books and movies in which someone goes from moribund old age to infanthood are fantasies, never to be realized in life.[7]

In equilibrium there is no such arrow of time. In equilibrium, order can increase only temporarily, through a random fluctuation. These excursions from equilibrium look, on average, the same when run forward or backward. If you took a movie of the motions of atoms in a gas in equilibrium and ran it backward, you wouldn't be able to tell which was the original version and which was reversed. Our universe is not like this.

The strong arrow of time we see in our universe requires explanation, because the fundamental laws of physics are time-symmetric. Any solution to their equations has a ghost-companion solution, which behaves just like the first but with the film run backward (with the added subtlety that left and right are switched and particles are replaced by their antiparticles). Thus the fundamental laws would not be violated if some people did age backward, or certain cups of coffee left on the counter got hotter, or shattered cups reassembled themselves spontaneously.

Why do these things never happen? And why do all these different asymmetries in time point the same way—toward increasing disorder? This is sometimes called the *problem of the arrow of time.*

There are actually several different arrows of time in our universe.

The universe is expanding and not contracting. We call this *the cosmological arrow of time.*

Small bits of the universe, left to themselves, tend to become more disordered in time (the spilt milk, the air equilibriating, and so on). This is called the *thermodynamic arrow of time.*

People, animals, and plants are born as infants, grow up, age, then die. This can be called the *biological arrow of time.*

We experience time flowing from the past into the future. We remember the past but not the future. This is the *experiential arrow of time.*

There is another—less apparent than the preceding arrows but nonetheless a major clue. Light moves from the past into the future. Hence, the light that reaches our eyes gives us a view of the world in the past, not the future. This is called the *electromagnetic arrow of time.*

Light waves are produced by the motion of electric charges. Wiggle a charge and light spreads out, always moving outward into the future, never into the past. This seems to apply to gravitational waves as well. So there is a *gravitational-wave arrow of time.*

Our universe apparently contains many black holes. A black hole is highly asymmetric in time. Anything can fall in, but all that comes out is Hawking thermal radiation. A black hole is a device for taking anything and turning it into a gas of photons in equilibrium. This irreversible process produces a lot of entropy.

But what about white holes? These hypothetical objects are solutions of general relativity gotten by reversing the direction of time in black holes. White holes behave in the opposite way from black holes. Nothing can fall into a white hole, but anything might come out of one. A white hole could look like the spontaneous appearance of a star, which is what you get if you take a movie of the collapse of a star into a black hole and run it backward. Astronomers have not seen anything that could be interpreted as a white hole.

Even if you consider just black holes, there is something strange about our universe. According to the equations of general relativity, it could well have started out filled with black holes. But, as noted in chapter 11, it seems there were none at all in the early universe. All the black holes we know about seem to have formed long after that, from the collapse of massive stars.

Why are there only black holes and no white holes? And why did the universe not start off filled with black holes? There seems to be a *black-hole arrow of time,* indicated by the absence of black holes in the universe's early history.

Could there be a galaxy on the other side of the universe where some of these arrows of time run backward? There's no evidence of this. We might live in a universe where some of the arrows of time are reversed from place to place, but apparently we don't. Why is this?

These distinct arrows of time are facts about our universe that require explanation. Any explanation offered for them rests on assumptions about the nature of time. The explanation offered by someone who believes time is emergent from a timeless world will differ from one offered by someone who believes time is fundamental and real.

Related to this is the question of whether or not the laws of physics are reversible. As noted in chapter 5, the fact that the laws of nature are time-reversible can be taken as evidence in favor of the view that time is not fundamental. How are we to explain the arrows of time if the laws of nature are reversible in time? The arrows of time each represent an asymmetry in time; how could they arise from time-symmetric laws?

The answer is that the laws act on initial conditions. The laws may be symmetric with regard to reversing the direction of time, but the initial conditions need not be. Initial conditions can evolve to final conditions that are easily distinguished from them. In fact this is the case: The initial conditions of our universe appear to have been finely tuned to produce a universe that is asymmetric in time.

Here's an example. The initial expansion rate of the universe, which is set by the initial conditions, seems to have maximized the production of galaxies and stars. Had it been much faster, the universe would have diluted too quickly for galaxies and stars to form. Had it been too slow, the universe might have collapsed directly to a final singularity before stars got a chance to form. The expansion rate was ideal for the production of lots of stars, and it is the stars that, by pouring hot photons into cold space for billions of years, keep the universe away from equilibrium and so explain the thermodynamic arrow of time.

The electromagnetic arrow of time can also be explained by time-asymmetric initial conditions.[8] At the universe's beginning, there were no electromagnetic waves. Light was produced only later, by the motion of matter. This explains why, when we look around, the images the light carries gives us information about the matter in the universe. If we just went by the laws of electromagnetism, it could be otherwise. The equations of electromagnetism allow the universe to begin with light traveling freely. That is, light would have formed directly in the Big Bang rather than being emitted from matter later on. In a universe like that, any images of objects that light carried away from matter would be swamped by the light coming straight from the Big Bang.

In such a world, we would not see stars and galaxies when we looked back with our telescopes. We might just see a random mess. Or, for that matter, the light formed in the Big Bang might carry images of things that were never there, like images of a garden with elephants munching on giant asparagus.

This is what the universe would look like if we took a movie of it at a time far in the future and ran it backward. In the far future, there will be lots of images traveling around—images of things that once existed. But if we run the movie backward in time, we see a universe filled with images of things that have yet to happen. Indeed, light carrying an image would flow into the event that the image represented and end there. The light we would see would tell us only about things yet to happen.

We don't live in such a universe, but if possible universes correspond to solutions of the laws of physics, we might. To explain why we see only things that happen or have happened and never anything yet to happen or that will never happen, we have to impose strict initial conditions. These forbid the universe to start with any light carrying images and flying freely about. This is a severely asymmetric condition to impose, but it is needed to explain the electromagnetic arrow of time.

A similar story holds for the gravitational-wave and black-hole arrows of time. If the fundamental laws are time-symmetric, then the whole burden of explaining why our universe is time-asymmetric falls on the choice of initial conditions. So you have to impose the condition that initially in the universe there are no gravitational waves moving freely, no initial or early black holes, and no white holes.

This point has been emphasized by Roger Penrose, and he has proposed a principle to explain it, which he calls the *Weyl curvature hypothesis.*[9] The Weyl curvature is a mathematical quantity that is nonzero whenever there is gravitational radiation or black or white holes. Penrose's principle is that at the initial singularity this quantity vanishes. He notes that this agrees with what we know about the early universe. It's a time-asymmetric condition, because it is certainly not true at late times in the universe. At late times, the universe has lots of gravitational waves and lots of black holes. Hence, Penrose argues, to explain the universe we see, this time-asymmetric condition has to be imposed on the choice of solution of the (time-symmetric) laws of general relativity.

That explaining our universe requires time-asymmetric initial conditions greatly weakens the argument that time is unreal because the laws of nature are time-symmetric. You cannot ignore the role of the initial conditions and declare that the past is like the future when, to agree even roughly with our universe, initial conditions have to be chosen that are very unlike those conditions that have evolved.[10]

The burden of explanation then falls on the question of how the initial conditions were chosen. But we know of no rational explanation for how they were chosen, so we reach a dead end, leaving a critical question about our universe unanswered.

There's another and much simpler option. We believe that our laws are approximations to some deeper law. What if that deeper law were time-asymmetric?

If the fundamental law is time-asymmetric, then so are most of its solutions.[11] There need then be no problem explaining why we never observe crazy things that would arise from running natural processes backward, because the time-reversal of a solution to the law will no longer be a solution. The mystery of why we see only images from the past and not the future is solved. The fact that the universe is highly time-asymmetric would be directly explained by the time asymmetry of the fundamental law. A time-asymmetric universe would no longer be improbable, it would be necessary.

This is what I understand Penrose had in mind when he proposed his Weyl curvature hypothesis. The difference between a physics near the initial singularity and a physics late in the universe would be forced on us by a quantum theory of gravity, which in Penrose's view should be a highly time-asymmetric theory. But a time-asymmetric theory is unnatural if time is emergent. If the fundamental theory contains no notion of time, we have no way to distinguish the past from the future. The extreme improbability of our universe would still require explanation.

A time-asymmetric theory is far more natural if time is fundamental. Indeed, nothing would be more natural than having a fundamental theory that distinguishes the past from the future, because the past and future are very different. Within a metaphysical framework in which time and the flow of moments from the past to the future are real, it is perfectly natural to have time-asymmetric laws governing a time-asymmetric universe. So the reality of time gains credence from these considerations, because it lets us avoid having to leave a vast improbability — the strong time-asymmetry of our universe — without explanation. Let's count this as another step in the discovery of time.

◆

Can we speak of the universe as being improbable?

Several times in this chapter I have referred to our universe or its initial conditions as improbable — for example, when I argued that it is improbable that a universe governed by time-symmetric laws have an arrow of time. But just what does it mean to assert that the universe is improbable? The universe is unique and happens just once. It is the only thing of its kind. Mustn't any property of it have probability?

To sort out this confusion, we need to know what we mean when we speak of some system being in an improbable configuration. Within the Newtonian paradigm, this makes sense, because the description refers to a subsystem of the universe, which can be one of many of its kind. But it clearly doesn't apply to the universe as a whole.

You might attempt to define the probability of our universe's having a particular property by supposing that the initial conditions were picked randomly from the configuration space. But we know that this supposition is false: We know that our universe wasn't produced by random choice, because of the many properties it has that would be extraordinarily unlikely to result from such a choice.

You could avoid this conundrum by imagining that there are a large number of universes. However, as we saw in Chapter 11, there are two kinds of multiverse theories: those in which our universe is atypical and hence improbable, like those generated by eternal inflation; and those exemplified by cosmological natural selection, which generates an ensemble of universes in which universes like ours are probable. As

I explained in Chapter 11, only in the latter kind are falsifiable predictions for doable observations possible; in the first class the anthropic principle must be used to select out our kinds of improbable universes and no predictions are possible by which the hypotheses underlying the scenario could be independently tested. We must conclude that whether there are many universes or just one, there is no empirical content to the statement that our universe is improbable.

But the whole science of thermodynamics is based on applying notions of probability to the microstate of a system. So it follows that we are committing the cosmological fallacy whenever we apply thermodynamics to discuss a property of the universe as a whole.[12] The only way to avoid the fallacy and the paradox of an improbable universe is to base our explanation for why the universe is complex and interesting on a time-asymmetric physics — a physics that makes a universe like ours inevitable rather than improbable.

This isn't the only instance of physicists reaching paradoxical conclusions by committing the fallacy of applying thermodynamics to the universe as a whole. Ludwig Boltzmann, who invented the statistical explanation for entropy and the second law of thermodynamics, seems to have been the first to propose an answer for why the universe is not in equilibrium. He didn't know about the expanding universe or the Big Bang; his conception of cosmology was of a universe that was eternal and static. The eternality of the universe was a great puzzle for him, because that meant it should already have reached equilibrium, since it had had an infinite amount of time to do so.

One reason he could think of for the universe not to be in equilibrium was that our solar system and the region surrounding it had relatively recently been the site of a very large fluctuation, in which the sun, planets, and surrounding stars formed spontaneously out of a gas in equilibrium. The entropy in our region was now increasing, as it found its way back to equilibrium. This was probably the best answer consistent with the picture of cosmology Boltzmann had near the end of the 19th century. But it is wrong. We know that now because we can see almost back to the Big Bang and out a corresponding 13 billion light-years, and we see no evidence for our region of the universe being a low-entropy fluctuation in a static world in equilibrium. We see instead a universe evolving in time, with structure on every scale developing as the universe expands.

Boltzmann could not have known this, but there's an argument he or his contemporaries might have used to put his explanation in doubt — stemming from the observation that the smaller the fluctuation, the more often it occurs in equilibrium. Hence the smaller the spatial region departing from equilibrium, the more probable it is.

Astronomers at the time of Boltzmann knew that the universe was at least tens of thousands of light-years across and contained many millions of stars. So if our region of space was the result of a fluctuation, it would have to be an extremely rare one — much rarer than other, smaller fluctuations that might have contained us. Consider a fluctuation that consists just of our solar system. We know we're not in one like that, because we would see nothing at night but infrared radiation coming from the gas in equilibrium surrounding us. But according to Boltzmann's assumptions, fluctuations like that should occur much more often in the equilibrium universe than is indicated by what we see — which is billions of stars, each as much out of equilibrium as our own solar system. It's considerably more likely that we would find ourselves in a solar-system-sized fluctuation than in a galaxy-sized one.[13]

We can go on from there. Most of the solar system is extraneous to our existence, so it's even more likely that we would find ourselves on Earth, with a hot spot in the sky, than in a solar system with the sun, seven other planets, comets, and the whole show. But this is only the beginning. All we really know is that we are thinking beings, perceiving ourselves to be in a world. But to produce a brain with memories and images would require much less of a fluctuation than one that produced a whole planet of living creatures orbiting a huge star. We can call a fluctuation that produces just one brain, complete with memories and experience of an imaginary world, a *Boltzmann brain.*

So there are a range of possibilities to explain our improbable existence as a fluctuation in Boltzmann's eternal equilibrium universe. We could be in a solar-system- or galaxy-sized fluctuation, one of trillions of living creatures on a planet, or we could be just a brain-sized fluctuation complete with images and memories. The latter takes much less information — that is, less negative entropy — so fluctuations of single brains occur much more often in the eternal universe than solar–system- or galaxy-sized fluctuations that contain whole populations of brains.

This is called the *Boltzmann brain paradox:* It implies that over an eternity of time there are vastly more brains in the universe which are formed from small fluctuations than brains arising in the slow process of evolution, requiring a fluctuation that lasts billions of years. So, as conscious beings, it is overwhelmingly probable that we are Boltzmann brains. But we know we are not such spontaneous brains — because if we were, it's more likely that our experience and memories would be incoherent than coherent. Nor is it likely that our brain would hold images of a vast universe of galaxies and stars around us. So Boltzmann's scenario turns out to be a classic *reductio ad absurdum.*

We should not be surprised, for we have committed the cosmological fallacy and it has led us to a paradoxical conclusion. The timeless view of physics based on the Newtonian paradigm has shown its impotence in the face of the most basic questions about the universe: Why is it interesting and why, indeed, is it so interesting that creatures like us can be here to marvel at it?

But if we embrace the reality of time, we make possible a time-asymmetric physics within which the universe can naturally evolve complexity and structure. And thus we avoid the paradox of an improbable universe.

17

Time Reborn from Heat and Light

I N   T H E   L A S T   C H A P T E R, we considered one of the greatest cosmological puzzles of all: why the universe is interesting and appears to be getting more and more interesting as time goes on. We saw that attempts to come to grips with this based on the timeless picture implied by the Newtonian paradigm led to two paradoxes: the claim that the unique universe is improbable and the Boltzmann brain paradox. In this chapter, I will explain how the principles of a new cosmological theory, enunciated in chapter 10, can lead to an understanding of why the universe is interesting while avoiding the paradoxes encountered in the last chapter.

We'll start with a simple question: Can the universe contain two identical moments of time?

The fact that there is an arrow of time means that every moment is unique. At least so far, the universe is different at different moments of time; these differences show in the properties of galaxies, say, or the relative abundances of the elements. The question is whether the progression of moments is accidental or reflects a deeper principle. In theories described within the Newtonian paradigm, the existence of an arrow of time appears accidental. In an eternal universe in equilibrium, we expect lots of pairs of identical or very similar moments.

But there is a deeper principle holding that no two moments of time can be identical. This is Leibniz's principle of the *identity of the indiscernibles,* which I described in chapter 10 as a consequence of his principle of sufficient reason. This principle holds that there cannot be two objects in the universe that are indistinguishable but distinct. This is just common sense. If objects are distinguished only by their observable properties, there cannot be two distinguishable objects that have exactly the same properties.

Leibniz's principle follows from the basic idea that physical properties of bodies are relational. What about two electrons, one of which is in an atom in the bedspread, the other on top of a mountain on the dark side of the moon? These are not identical particles, because their location is one of their properties. From a relational point of view, we can say that they're distinguishable by having distinguishable surroundings.[1]

There is no absolute space, so there's no way to ask what's happening at a particular point without giving instructions for how to recognize that point. So we cannot locate an object at a point unless we have some way to specify that place. One way to recognize where you are is by noting what's unique about the view from there. Suppose someone were to claim that two objects in space have exactly the same properties and exactly the same surroundings. This means that no matter how far from the two objects you explored, you would discover the same organization of everything else in space. If this weird situation existed, there would be no way to tell an observer how to distinguish one object from another.

So to ask the world to contain two identical objects is to demand the impossible. It means that there must be two identical places in the universe — locations from which the view of the universe is exactly the same. The universe as a whole is then greatly shaped by the seemingly simple request that it not contain two identical objects.[2]

The same argument applies to events in spacetime. The principle of the identity of the indiscernibles requires that there cannot be two events in spacetime that have exactly the same observable properties. Nor can there be two moments of time that are identical.

When we look out at the night sky, we see the universe from the vantage point of a particular place at a particular moment of time. The view includes all the photons arriving to us from near and far. If physics is relational, then these photons make up the intrinsic reality of that particular event — that is, your looking up at the night sky at that particular place and time. The principle of the identity of the indiscernibles then says that the view of the universe an observer could see from each event in the history of the universe is unique. Suppose aliens kidnapped you while you slept and took you on a trip in their time machine. In principle, if you were to awaken and find yourself in some distant galaxy far from home, you could tell exactly where you were in the universe by making a map of what you see when you look around. You could further tell exactly when in the universe you had been transported to.

This implies that our universe can have no exact symmetries. In fact it doesn't, as discussed in chapter 10. Whereas symmetries are helpful for the analysis of models of small parts of the universe, all the symmetries so far posited by physicists have turned out to be approximate or broken.

According to the principle of the identity of the indiscernibles, our universe is one where every moment of time, and every place at every moment, is uniquely distinguishable from any other. No moment ever repeats. Looked at in enough detail, every event in the universe is unique. In such a universe, there is never a complete realization of the conditions needed to make sense of the Newtonian paradigm. That method, as noted, requires that we can repeat experiments many times to check their replicability as well as to distinguish the effect of a general law from effects of changing the initial conditions. This can be achieved approximately but never exactly, because the more detail we note, the more apparent it is that no event or experiment can be an exact copy of another.

It will help to have a name for hypothetical universes in which every moment of time and each and every event is unique. We'll call a universe that satisfies the principle of the identity of the indiscernibles a *Leibnizian universe.*

This is in stark contrast to the universe envisioned by Ludwig Boltzmann. In that vision of cosmology, most of the universe's history is dominated by periods of thermal equilibrium, where entropy is maximized and there is no structure or organization. These long deathlike periods are punctuated by relatively short periods in which structure and organization arise due to a statistical fluctuation — and then dissipate, due to the tendency of entropy to increase. We can call such a world a *Boltzmannian universe.*

The question on which the future depends is, Do we live in a Boltzmannian universe or a Leibnizian universe? In a Leibnizian universe, time is real, in the sense that no moment of time is like any other. In a Boltzmannian universe, there are lots of moments that recur — if not precisely, then to any degree of precision you might want. In an approximate sense, most moments of a Boltzmannian universe are like all the others, because all moments in equilibrium are roughly the same. The bulk quantities, such as temperature and density, that measure averages are uniform. True, the atoms fluctuate around those averages but almost never enough to amount to macroscopic levels of structure and organization. In a Boltzmannian universe, if you wait long enough the universe will come as close as you like to repeating any configuration. On average, these near recurrences are separated by the Poincaré recurrence time. But if time is eternal, each moment repeats an infinite number of times.

A Leibnizian universe is just the opposite: By definition, no moment ever recurs in a Leibnizian universe. A universe cannot be both Boltzmannian and Leibnizian. So which is ours?

If time is real, it should be impossible to have two different but identical moments of time. Time is fully real only in a Leibnizian universe. A Leibnizian universe will be full of complexity that generates a bountiful array of unique patterns and structures. And it will be ever changing, to ensure that every moment can be distinguished from every other by the structures and patterns present then. As indeed is our universe.

❖

It is good to know that our universe appears to satisfy a grand principle such as the principle of the identity of the indiscernibles, but this does not take away all the mystery. For principles do not act on matter, laws do. We need to know how the principle acts through the laws to ensure its satisfaction. To some extent, we know the answer — which has to do with gravity's twisted relation to thermodynamics.

One component of our present Leibnizian universe is nearly in thermal equilibrium; this is the cosmic microwave background — but the CMB, we know, is a relic of the early universe, having arisen about 400,000 years after the Big Bang. Certainly equilibrium rules in the vast regions of interstellar and intergalactic space. Much of the universe, though, is far from equilibrium. The most common objects in our universe are stars, and these are not in equilibrium with their surroundings. A star is always in a dynamical balance between the energy generated by nuclear reactions in its core, which would blow it up, and gravity, which would collapse it. It will reach what Boltzmann would call equilibrium only when its nuclear fuel runs out and it settles down as a white dwarf, a neutron star, or a black hole (except that if it's a black hole, it may become the engine of a system that accretes matter and then accelerates it outward). Such systems are not in equilibrium, however; they are dynamical steady states.

A star can be characterized as a system driven far from equilibrium by a steady flow of energy through it. The energy comes from both nuclear and gravitational potential energy, which is slowly converted into starlight in a range of frequencies. The starlight then illuminates the surfaces of planets, like ours, driving them into far-from-equilibrium states of their own.

This is an example of a general principle[3]: *Flows of energy through open systems tend to drive them to states of higher organization.* ("Open systems," recall, are any bounded systems that can exchange energy with their surroundings.) We can call this the principle of *driven self-organization.* If the principle of sufficient reason is the paramount explanatory principle in nature and the identity of the indiscernibles her prince, the principle of driven self-organization is the good angel who does the detailed work in myriads of stars and galaxies to ensure a diverse, complex universe.

Fill a pot with water and put it on the stove. The system (the pot and the water in it) is an open one, because energy is being slowly introduced at the bottom, which heats the water before passing out through the surface and into the air. To make the point simplest, let's put a lid on the pot, to prevent the water from escaping even when it turns to steam. After a while, the water comes to a steady state in which neither its temperature nor its density is uniform. The temperature of the water is hottest at the bottom and decreases toward the surface; the density behaves in the opposite manner. The energy passing through the water has displaced the water from equilibrium. Soon a structure starts to appear: convection cycles, in which the water moves in an ordered way in columns. The cycles are driven by the heat input from the bottom. The water is heated, expands, and hence moves upward as a column of rising water. At the surface, it gives up some of its heat, becomes denser than its surroundings, and sinks, creating a column of falling water. Because water cannot rise and fall in the same space, structure is created, as the rising and falling columns segregate themselves.

The steady flow of energy through a system can result in complex patterns and structures, evidence that these systems are far from thermodynamic equilibrium. Another example is the wind-created rippling on sand dunes. At the other end of the spectrum of complexity is life. Both, and many things in between, are the result of the steady energy flow through a system. This means, among other things, that complex self-organized systems are never isolated.

These flows produce systems that are robustly Leibnizian. Living things tend to come in many copies, but each is distinguishable from the others. And the further you go up the ladder of complexity, the more distinguishable individuals are from one another.

There is much beautiful science down that road. The point is that, as noted in the preceding chapter, you cannot apply the second law of thermodynamics except to an isolated system, enclosed in a box that prevents matter and energy from being exchanged with the outside. No living system is an isolated system. We all ride flows of matter and energy — flows driven ultimately by the energy from the sun. Once enclosed in a box (in a prefiguration of our eventual interment), we die.

Thus Aristotle was right when he understood that the earthly realm is kept from equilibrium by the flow of energy through it. Insufficient appreciation of this idea has led some scientists and philosophers to see a conflict between the second law of thermodynamics and the fact that natural selection produces increasingly improbable structures. There is no contradiction, because the law of increasing entropy does not apply to the biosphere, which is not an isolated system. Indeed, natural selection is a mechanism of self-organization that may spontaneously arise as a consequence of the tendency of externally driven systems to organize themselves.

Within the context of self-organizing systems we can understand better what features make a system complex. Highly complex systems cannot be in equilibrium, because order is not random, so high entropy and high complexity cannot coexist. Describing a system as complex does not just mean that it has low entropy. A row of atoms sitting in a line has low entropy but is hardly complex. A better characterization of complexity, invented by Julian Barbour and myself, is what we call variety: a system has high variety if every pair of its subsystems can be distinguished from each other by giving a minimal amount of information about how they are connected or related to the whole.[4] A city has high variety because you can easily tell from looking around which corner you are on. Such conditions arise in nature in systems far from equilibrium as a result of processes of self-organization.

A ubiquitous feature of such self-organizing systems is that they are stabilized by feedback mechanisms. Any living thing is an intricate network of feedback processes that regulate, channel, and stabilize the flows of energy and material through it. Feedback can be positive, which means that it accelerates the production of something (like the screeching of a microphone when it gets too close to a speaker). Negative feedback acts to damp a signal, as in the thermostat that turns on your furnace when the house is too cold and turns it off when it's too warm.

Patterns in space and time are formed when different feedback mechanisms compete to control a system. When a positive-feedback mechanism competes with a negative-feedback mechanism but they act on different scales, you may get patterns in space. This basic mechanism of biological self-organization, discovered by Alan Turing,[5] acts to produce the patterns in an embryo that mark out the parts of the body it will become. Later on it may act again, to produce patterns on the skin of a cat, for instance, or the wings of a butterfly.

What do we see when we look beyond the scale of stars and solar systems? Stars are organized into galaxies, because that's where they're made. Galaxies are themselves far from thermodynamic equilibrium. Our own Milky Way is a typical spiral galaxy. It contains not only stars but also vast interstellar clouds of gas and dust, out of which stars form. The gas slowly accretes onto the disk of the galaxy from outside; this is one of the drivers of change in a galaxy. The dust is produced by stars and injected into the galactic disk when stars blow up at the end of their lives as supernovas. The gas and dust exist in different phases; some is very hot, and some is condensed in very cold clouds. The processes of self-organization in a galaxy are driven by starlight — energy flows coming from the stars. From time to time, a massive star explodes in a supernova, and that also pours a lot of energy and matter into the galaxy. We also see structure above the scale of galaxies, which are organized into clusters and sheets separated by voids. These patterns are believed to be formed by dark matter and held together by its interactions.

So our present universe is characterized by structure and complexity on a wide range of scales, from the organization of molecules in living cells to the organization of galaxies into clusters. There is a hierarchy of self-organizing systems, driven by energy flows and stabilized and shaped by feedback processes. This is a universe that is far more Leibnizian than Boltzmannian.

What do we see when we look back? We see a universe evolving from less to more structured, from equilibrium to complexity.

There is good reason to believe that the matter and radiation in the early universe was nearly in thermal equilibrium. The matter and radiation were in a hot state, with a remarkably uniform temperature, which increases as we go further back in time. Before the era of decoupling (the separation of photons from matter 400,000 years after the Big Bang), the matter was in equilibrium with the radiation — an equilibrium that was, as far as we know, disturbed only by random density fluctuations. All the structure and complexity we see today formed after matter and radiation decoupled. The initial structures were seeded by the small random density fluctuations, and these structures grew as the universe expanded. Galaxies formed, then stars, then life.

This is certainly not the picture a naïve application of the second law of thermodynamics would suggest. The second law says that isolated systems increase their randomness, becoming more disordered and less complex and structured as time moves forward. This is the opposite of what we see happening in the history of our universe, in which complexity increases as structures form on many scales, with the most intricate structures being the most recent.

Evolving complexity means time. There has never been a static complex system. The big lesson is that our universe has a history, and it is a history of increasing complexity with time. The universe is not only not Boltzmannian, it is becoming less and less Boltzmannian as time goes on.

This doesn't abrogate the second law of thermodynamics. The second law applies to isolated systems, and these come to equilibrium over time. Moreover, the formation of complexity is actually compatible with an increase in entropy, as long as the entropy increase and the growing complexity are occurring in different places. The Earth's biosphere has been organizing itself for nearly 4 billion years, ever since the origins of life on our planet. This increasing organization is driven by the flow of energy from the sun, arriving as photons of mostly visible light, which is captured by photosynthesis in plants. Photosynthesis captures the photons' energy in chemical bonds. In this form, the energy can catalyze chemical reactions that might, for example, form a protein molecule. The energy eventually passes through the biosphere, escapes as heat, and ultimately is radiated as infrared photons into the sky and beyond. A photon's next stop might be to heat a grain of dust in orbit around the sun.

A single quantum of energy may have catalyzed the formation of a complex molecule and hence lowered the entropy of the biosphere, but when it is radiated as infrared light out into space, that increases the entropy of the solar system as a whole. As long as the increase in entropy caused by heating a dust grain somewhere in space is greater than the decrease of entropy caused by forming a molecular bond, the long-term outcome is in agreement with the second law.

So if we consider the solar system as an isolated system, the fact that parts of it are undergoing self-organization is compatible with an overall increase of its entropy. The system as a whole is trying to come to equilibrium and will increase its entropy where it can. The second law is doing its best to drive the solar system to equilibrium, but as long as there's a big star radiating hot photons into cold space, that equilibrium is postponed. While it's postponed, molecules can ride the energy flow to greater and greater states of organization and complexity. And stars burn for billions of years, so there's lots of time for complexity to proliferate. The existence of stars has much to do with why the universe is far from equilibrium almost 14 billion years after its formation.

◈

But why are there stars? If the universe must tend toward entropy and disorder, how is it that stars, which drive the universe away from equilibrium, are ubiquitous? To put this another way: If the universe is to be Leibnizian, something like stars must exist. What features of the laws of nature guarantee that they do?

The physics of stars relies on two unusual features of the laws of nature. The first is incredibly fine tuning in the parameters that govern physics. These fine tunings include the masses of the elementary particles and the strengths of the four forces. They make nuclear fusion possible, so the hydrogen gas comprising a star does not behave as it would in the absence of nuclear forces. Rather than just moving around randomly, the hydrogen atoms jammed together at the center of a star can interact in a new way. They fuse to make helium and a few other light elements. It's as if you were trapped in a cell, day after day, in the same boring equilibrium. Every hour is like every other. Then all of a sudden a door opens where there was none before, and you escape into a whole new world. The laws of thermodynamics applied to generic atoms would never predict nuclear fusion and the possibilities it gives rise to.

The second unusual feature has to do with the behavior of systems held together by the force of gravity. Very simply, gravity subverts our naïve ideas about thermodynamics.

An everyday observation, which is also a consequence of the second law of thermodynamics, is that heat flows from hotter bodies to colder bodies. Ice melts. Water on the stove boils. Heat stops flowing when the temperature of the two bodies is the same; they have reached the state of equilibrium. Normally when we take energy out of a body, its temperature goes down, and when we put energy into a body, it heats up. So when heat flows from a hotter body to a colder body, the latter heats up and the former cools down. This goes on until they're at the same temperature. This is why the air in a room is at a single temperature. If it weren't, energy would flow from the warmer side to the cooler until they reached a common temperature.

This behavior makes the system in equilibrium stable against the effects of small fluctuations. Suppose, by a small fluctuation, one side of a room became a bit warmer than the other. Energy would flow from the warm side, cooling it, to the cooler side, warming it, so that soon the temperature is uniform again. Most systems work in this intuitive way. But not all.

Imagine there's a gas that works the other way, cooling down when you add energy to it and heating up when you take energy away. This may seem counterintuitive, but there are such gases. They have to be unstable. Suppose you start off with all of this kind of gas in a room at the same temperature. A little fluctuation moves a bit of energy from the left side to the right. Then the left side heats up, while the right side cools down. This causes more energy to flow from the left side, the hot side, to the cold side. As it does, the left side won't cool down; rather, it gets even hotter. And as more energy flows into the cool right side, that side gets even cooler. Soon you have a runaway instability, in which the two sides of the room are driven to continually increase their temperature difference.

Now let's look at just the hot side and repeat the scenario. Suppose another fluctuation appears, cooling the center of the hot side a bit. The same phenomenon acts as positive feedback to further cool the center and further heat the region around it. As time goes on, the little fluctuation grows into a feature. This can happen again and again. Soon you have a complex pattern of cold and hot regions.

A system that works this way naturally drives itself to form complex patterns. It's hard to predict where such systems will end up, because there are a huge number of heterogeneous, patterned configurations it might evolve toward. We call these *anti-thermodynamic systems.* The second law still operates in them, but because putting energy into a region cools it down, the state in which the gas is uniformly distributed is highly unstable.

Systems held together by gravity behave in this crazy way. Stars, solar systems, galaxies, and black holes are all anti-thermodynamic. They cool down when you put energy into them. This means that all these systems are unstable. The instabilities drive them away from uniformity and stimulate the formation of patterns in space and time.

This has a lot to do with why the universe is not in equilibrium 13.7 billion years after its origin. The increasing structure and complexity that characterize the universe's history are largely explained by the fact that the gravitationally-bound systems filling it, from clusters of galaxies to stars, are anti-thermodynamic.

It's easy to understand why such systems are anti-thermodynamic. Two basic features differentiate gravity from the other forces: The gravitational force is (1) long-range and (2) universally attractive. Consider a planet in orbit around a star. If you put energy in, it will move to an orbit farther from the star, where it moves slower. So putting energy in decreases the speed of the planet, and this lowers the system's temperature — because temperature is just the average speed of things in the system. Conversely, if you take energy out of the solar system, the planet must respond by falling closer to the star, where it moves faster. Hence, taking energy out heats up the system.

We can compare this with the behavior of an atom, which is held together by the electric force between charges. Like gravity, the electric force acts over long distances, but it differs by being attractive only between opposite charges. A positively-charged proton will attract a negatively-charged electron, but once the electron is bound to the proton the resulting atom has no net charge. The force is said to saturate, and the atom does not attract any other particles to it. A solar system works the opposite way, because when a star attracts some planets, the resulting system is even more attractive to passing bodies than the star alone would have been. So here's another instability — a gravitationally-bound system will attract still more bodies to it.

This anti-thermodynamic behavior manifests itself in the devolution of star clusters. If a star cluster were to act thermodynamically, it would reach equilibrium — in this case, a state in which all its stars had the same average speed and stayed clustered forever. Instead, what happens is that a star cluster slowly dissipates. This happens in an interesting way. Every once in a while a star comes close to a double star — that is, two stars in orbit around each other. A close approach can result in a narrower orbit for the double star. This orbital shrinking releases energy, which is imparted to the third star. The third star now has enough energy to escape the cluster, and it begins a journey off into space. After a long time, little is left of the star cluster except some double stars in close orbits and a cloud of fast-moving stars streaming away from the cluster.

This does not contradict the second law, only a naïve interpretation of it. The law that entropy should usually increase just codifies the truism that the more ways there are for something to happen, the more likely it is that it will. Normal thermodynamic systems end up in the single, boring state of uniform equilibrium; gravitationally-bound, anti-thermodynamic systems end up in one of a large number of highly heterogeneous states.

So the fact that our universe is interesting has a threefold explanation: The *principle of driven self-organization* acts over a myriad of subsystems and scales, from the molecular to the galactic, evolving them to states of ever increasing complexity. The engines driving that process are the stars, which exist because of a combination of the *fine tuning of the fundamental laws* and the *anti-thermodynamic nature of gravity.* But these forces can produce a universe filled with stars and galaxies only if the initial conditions of the universe are strongly time-asymmetric.

All this can be framed and to some extent understood within the Newtonian paradigm. But if we continue to think within that paradigm, the world's organization seems to rest on vast improbabilities — the extreme specialness of the choices of laws and initial conditions. The sad conclusion is that the only kind of universe that appears natural from the timeless perspective of the Newtonian paradigm is a dead universe in equilibrium, obviously not the kind we live in. But from the perspective of the reality of time, it is entirely natural that the universe and its fundamental laws be asymmetric in time, with a strong arrow of time that encompasses increases of entropy for isolated systems together with continual growth of structure and complexity.

18

Infinite Space or Infinite Time?

W E HAVE SEEN THAT by embracing the reality of time we can comprehend why the universe is full of structure and complexity. But how long can it stay complex and structured? Can equilibrium be held off forever? Maybe we're just in a bubble of complexity in a much larger equilibrium universe.

This brings us to the most speculative subjects in modern cosmology: the very far away and the far future.

There is no more romantic notion than infinity, but in science the concept can easily lead to confusion. Imagine that the universe is infinite in spatial extent. Imagine also that the same laws hold throughout but that the initial conditions were chosen randomly. This is a picture of the ultimate Boltzmannian universe. Almost all of the infinite universe is in thermodynamic equilibrium; anything interesting that happens is a consequence of a fluctuation. But anything that can happen in a fluctuation will happen somewhere, and if there is an infinite quantity of "somewheres" available, each fluctuation, no matter how improbable, will happen an infinite number of times.[1]

So our observable universe could be just a big statistical fluctuation.

If the universe really is infinite, then our observable universe, which is a region about 93 billion light-years across, will be repeated an infinite number of times throughout the infinity of space. So if the universe is infinite and Boltzmannian, we exist, just as we are, and act, just as we are acting, an infinite number of times.

This certainly violates the Leibnizian principle that there can be no two places in the universe that are identical.

But not only that. Imagine, any way you like, that today could have been different. I might not have been born. Or you married your first boyfriend. Someone got drunk a year ago, didn't heed my friends' advice, drove home, and struck and killed a child on the way. Your cousin was accidentally switched at birth, brought up by an abusive family, and became a mass murderer. A race of intelligent dinosaurs evolved, solved their climate-change problem, and still dominate the planet, so that mammals never took over. All these are things that might have happened, bringing us to a different present configuration of the universe. Each such present configuration is a possible way that the atoms in our neighborhood might be arranged. So each occurs an infinite number of times in the infinity of space.

This to me is a horrifying prospect. It raises ethical issues, for why should I care about the consequences of the choices I make, if all the other choices are made by other versions of me in other regions of the infinite universe? I can choose to nurture my child in this world, but shouldn't I care also for the children in other worlds who suffer because of the bad decisions made by my other selves?

In addition to the these ethical issues, there are issues pertaining to the usefulness of science. If the real fact of the world is that anything that might happen does, then the scope for explanation is much reduced. Leibniz's principle of sufficient reason demands that there be a rational reason for every case where the universe is one way but might have been another. But if the universe is every possible way, there is nothing to explain. Science may give us insight into local conditions, but ultimately it is a fruitless exercise, because the true law will simply be that anything that might happen is happening an infinite number of times, right now. This is a kind of *reductio ad absurdum* of the Newtonian paradigm extended to cosmology — and another instance of the cosmological fallacy. I call it the *infinite Boltzmannian tragedy.*

One reason it's a tragedy is that the predictive power of physics is greatly reduced, because probabilities don't mean what you think they mean. Suppose you're doing an experiment for which quantum mechanics predicts that outcome A is 99-percent probable and outcome B is 1-percent probable. Suppose you do the experiment 1,000 times. Then you can expect that roughly 990 of those times A will result. You would feel safe betting on A, because you can reasonably expect roughly 99 outcomes of A for every 1 of B. You'd have a good chance of confirming the prediction of quantum mechanics. But in an infinite universe there are an infinite number of copies of you doing the experiment. An infinite number of these copies have you observe outcome A. But there are also an infinite number of copies of you observing outcome B. So the prediction of quantum mechanics that one outcome is 99 times more frequent than the other is not verifiable in an infinite universe.

This is called the *measure problem* in quantum cosmology. After having read and listened to the bright people working on it, my view is that it's not solvable. I prefer to take the fact that quantum mechanics works as evidence that we live in a finite universe containing only a single copy of me.

We can avoid the implications of the tragedy of an infinite universe by denying that the universe is infinite in space. Whereas, of course, we can't see past a certain distance, it seems plausible and sensible to me to hypothesize that the universe is finite in spatial extent — as in Einstein's proposal that it is finite but unbounded. This means that the universe has an overall topology of a closed surface, like a sphere or a doughnut (that is, a torus).

This proposal does not contradict our observations. Which topology is correct depends on the average curvature of space. If the curvature is positive, like a sphere, then there is only one possibility, which is the three-dimensional analogue of a sphere's two-dimensional topology. If the average curvature of space is flat, like a plane, then there is also one choice for a finite universe, which is the three-dimensional analogue of a doughnut's two-dimensional topology. If the curvature is negative, like a saddle, then there are an infinite number of possibilities for its topology. These are too complex to be described here, and cataloguing them was a triumph of late-20th-century mathematics.

Einstein's proposal is a hypothesis that could be confirmed. If the universe is closed, and small enough, then light should go all the way around, and we should see faraway galaxies in multiple images. This has been looked for and, so far, hasn't been found.

There is, however, a strong reason to prefer that a cosmological theory be modeled by a spacetime that is spatially closed. If the universe is not spatially closed, then it must be infinite in spatial extent. This means, counterintuitively, that there is a boundary to space. This boundary is infinitely far away, but nonetheless it's a boundary, which information could pass through.[2] Consequently, a universe that is spatially infinite cannot be considered a self-contained system. It must be considered a part of a larger system that includes whatever information is coming in from the boundary.

If the boundary were a finite distance away, you could imagine that there was still more space outside it. The information about the boundary would be explicable in terms of what is coming in from the world beyond the boundary.[3]

But the boundary at infinity does not allow us to imagine a world beyond. We are simply required to specify information about what is coming in and going out there, but the choice is entirely arbitrary. There can be no further explanation for the information coming into the universe from the infinite boundary: A choice must be made, and the choice is arbitrary. Hence, we have to concede that nothing can be explained in any model of a universe that has an infinite boundary. The principle of explanatory closure is violated and with it the principle of sufficient reason.

There are technical subtleties in this argument that I won't mention here. But the argument is a crucial one, which, as far as I can tell, is ignored by cosmologists who speculate that the universe is spatially infinite. I see no way to escape the conclusion that any model of a universe must be spatially closed, without boundary.

So there is nothing infinitely far away and no infinite spaces to contend with. Now let's turn our attention from infinite distance to the infinite future.

◆

The literature of cosmologists is filled with an anxiety about the future. If the universe is, so far, more Leibnizian than Boltzmannian, might this be so only temporarily? Perhaps in the long term not only will we all die but so will the universe.

The restriction to spatially finite universes gets us out of many of the tragedies and paradoxes of an infinite Boltzmannian universe. However, it does not get us out of all of them. The spatially finite and closed universe still may live for an infinite time, and if it never contracts it will expand forever. There is then an infinite amount of time available for it to reach thermal equilibrium. If it does, and no matter how long that takes, there will remain an infinite amount of time as well as a continually growing amount of space for fluctuations to create improbable structures. Consequently we can argue here, too, that anything that can happen will eventually happen an infinite number of times. This leads again to the Boltzmann brain paradox. If the principles of sufficient reason and the identity of the indiscernibles are to be satisfied, the universe must by some means avoid ending up in such a paradoxical state. These principles limit the options for the possible future fate of the universe.

There's a small scientific literature attempting to chart what will happen far in the universe's future. It's all speculative, because to reason about the far future, you have to make some big assumptions. One is that the laws of nature must never change, for if they did our predictive ability would be stymied. And no undiscovered phenomena must exist that could change the course of the universe's history. There might, for example, be some force so weak it has yet to be detected but nonetheless comes into play over vast distances and over times much longer than the universe's present age. This is possible and has been contemplated. But it hampers any prediction from current knowledge. There must also be no other surprises in store, such as walls of cosmic bubbles coming at us at the speed of light from beyond our present horizon.

Assuming that the well-established laws and phenomena are all there is, we can reliably deduce the following:

Eventually the galaxies will stop making stars. Galaxies are giant systems for turning hydrogen into stars. They're not very efficient; a typical spiral galaxy makes around a star a year. After almost 14 billion years, most of the universe is still primordial hydrogen and helium. Yet there is only so much hydrogen, so at the very least there can be only a finite number of stars. Even if all the hydrogen is eventually processed into stars, there will be a last star made. And this is just an upper limit; most likely the non-equilibrium processes that drive star formation will peter out long before all the hydrogen has been made into stars.

The last stars will burn out. Stars have a finite lifetime. The massive ones live a few million years and die dramatically as supernovas. Most live many billions of years and end with a fizzle as white dwarfs. There will be a time after the last star has died.

Then what?

Once the last stars have died, the universe is filled with matter, dark matter, radiation, and dark energy. What happens to the universe in the long term depends mostly on the component we know the least about: dark energy.

Dark energy is energy associated with empty space. It has been observed to make up about 73 percent of the mass-energy of the universe. Its nature is so far unknown, but its effect on the motion of distant galaxies has been observed. In particular, dark energy is invoked to explain the recently discovered acceleration of the universal expansion.

Apart from that, we know nothing about it. It could simply be a cosmological constant or it could be some exotic form of energy with a constant density. Although the density of the dark energy appears to be roughly constant, we don't know if that's really so or if it's just changing more slowly than observations have so far detected. The universe's future will be very different depending on whether the density of dark energy remains constant or not.

Let's look first at the scenario in which the dark energy retains its density as the universe expands. If it has a constant density, it behaves just like Einstein's cosmological constant. It does not decrease as the universe continues to expand. Everything else — all the matter and all the radiation — is diluted as the universe expands, and the total energy density from those sources decreases steadily. After a few tens of billions of years, everything is negligible except for the energy density associated with the cosmological constant.

Because this is such a simple case, we have a pretty good idea of what happens. A consequence of the exponential expansion is that galaxy clusters separate so quickly that soon they can no longer see one another. Photons leaving one cluster and going at the speed of light do not move fast enough to catch up with other clusters. Observers in each cluster are surrounded by a horizon beyond which their neighbors have vanished. Each cluster is then an isolated system. The interior of each horizon is thus a kind of box, delimiting a subsystem from the rest of the universe. So the methods of physics in a box apply to each — which means we can employ the methods of thermodynamics to reason about them.

At this point in the story, a new effect of quantum mechanics enters, causing the interior of each horizon to fill with a gas of photons in thermal equilibrium — a kind of fog created by processes analogous to those that create the Hawking black-hole radiation. It's called the *horizon radiation.* Its temperature is extremely low, and so is its density, but they remain constant as the universe expands. Meanwhile, everything else, including the matter and the CMB is becoming more and more dilute, so after enough time has passed, what fills the universe is this horizon radiation. The universe has come to equilibrium.

This state of equilibrium persists forever. There is no avoiding ending as an eternal Boltzmannian universe. There will be fluctuations and recurrences, of course, and occasionally one or another of a configuration of the universe will exactly recur — including that of the Boltzmann brain paradox I described in chapter 16 as the final *reductio ad absurdum* of the Newtonian paradigm. According to this scenario, the apparent complexity of our universe until now is just the briefest flash before the universe settles down into its eternal equilibrium.

We know with near certainty that we are not Boltzmann brains, because (as noted in chapter 16) if we were, we would probably not see a vast and ordered universe around us. The fact that we are *not* Boltzmann brains means that this scenario for the future of our universe is false. The principle of sufficient reason, acting through its surrogate, the principle of the identity of the indiscernibles, also requires the scenario to be false. The question is, How is it to be avoided?

The simplest way to avoid the eternal dead universe would be if the universe had enough density of matter to stop the expansion and cause it to collapse. Matter attracts matter gravitationally, and this slows the expansion, so if there is enough matter the universe will collapse to a final singularity. Or perhaps quantum effects will stop the collapse and "bounce" the universe, turning contraction into expansion leading to a new universe. But there doesn't seem to be enough matter to reverse the expansion, let alone counteract the tendency of dark energy to accelerate it.

The next simplest way to avoid an infinite dead future is if the cosmological constant is not actually a constant. While we have evidence that the dark energy — which is for all intents and purposes the cosmological constant — is not changing on scales of the present age of the universe, we have no evidence that it will not change in the long run. This change could be due to a deeper law, one that acts so slowly that its effects are perceptible only on long time scales. Or the change could just be an effect of the general tendency for laws to evolve. Indeed, the principle of no unreciprocated action suggests that the cosmological constant should be influenced by the universe on which it acts so decisively.

The cosmological constant could decay to zero. If it does, the expansion slows but most likely doesn't reverse. The universe might be eternal but static; this at least avoids the Boltzmann brain paradox.

Whether the universe with no cosmological constant expands for-ever or collapses depends ultimately on the initial conditions. If the energy in the expansion is ultimately sufficient to overcome the mu-tual gravitational attraction of everything in the universe, it will never collapse. But even if the universe is eternal, there's ample opportunity for rebirth, since each black hole, as a result of the elimination of its singularity, may lead to the birth of a baby universe. As noted in chap-ter 11, there is good theoretical evidence that this must happen.

If this is the case, then our universe, which is still far from dead, has already had at least a billion billion progeny. These new universes will each give birth to further progeny. The fact that each universe may die at some point, after having spawned so many others, thus seems in-consequential.

There are also possibilities for rebirth that involve the entire uni-verse instead of just its black holes. This is the hypothesis investigated in a class of cosmological models called cyclic models. One species of cyclic models, invented by Paul Steinhardt of Princeton University and Neil Turok of Perimeter Institute, accomplishes this by presuming that the cosmological constant decreases to zero and then keeps going to strongly negative values.[4] For reasons I won't explain here, this causes a dramatic collapse of the entire universe. However, they argue that this collapse is followed by a bounce and a re-expansion. This bounce could be due to the effects of quantum gravity, or the ultimate singu-larity might be evaded by the dark energy's extreme value.

The theoretical evidence that cosmological final singularities bounce due to quantum effects, leading to a re-expansion of the uni-verse, is even stronger than in the case of black-hole singularities.[5] Within loop quantum gravity, several models of quantum effects near cosmological singularities have been studied, and the result is that the bounce is a universal phenomenon. It must, however, be cautioned that these are only models, and so far they make drastic assumptions. The key assumption is that the universe is spatially homogeneous.

What we are surest of is that highly uniform regions of universes — regions without gravitational waves or black holes — bounce to give rise to new universes.

In the worst case, regions that are highly inhomogeneous will not bounce. They just collapse to singularities, where time stops. However, even this bad case has a silver lining, for it would provide a selection principle to determine which regions of the universe bounce and reproduce themselves. If only the more homogeneous regions bounce, then the beginnings of new universes, just after the bounce, will also be highly homogeneous.[6] This gives a prediction: At very early times, just after the bounce, the universe is highly homogeneous — there are no black or white holes and no gravitational waves, just as we see in our universe.

But for the bouncing-universe scenario to be science, there must be at least one more prediction by means of which the hypothesis could be tested. There are at least two, which have to do with the spectrum of fluctuations in the CMB. The cyclic scenario offers an explanation for those fluctuations that does not require the short period of extreme inflation that has often been taken as their cause. The spectrum of fluctuations we have so far seen is reproduced, but there are two differences between the predictions of the cyclic models and those of inflation, and these predictions can be tested in present and near-future experiments. One test is whether gravitational waves will be observed in the CMB; inflation says yes and the cyclic models say no. The cyclic models also predict that the CMB radiation is not completely random — in technical language, they predict non-Gaussianity.

The cyclic models are examples of how considering time as fundamental — in the sense that time did not begin at the Big Bang but existed before it — leads to a cosmology that is more predictive. Another example is that of theories in which the speed of light is hypothesized to have been different — in fact, much faster — in the very early universe. These so-called *variable-speed-of-light theories* pick out a preferred notion of time in a way that violates the principles of relativity theory. Consequently they're not popular, but they do hold promise for explaining the CMB fluctuations without inflation.

Roger Penrose has proposed another scenario for getting the universe to give rise to a new universe.[7] Roughly speaking, he accepts the scenario of an eternal Boltzmannian universe with a fixed cosmological constant and then asks what happens after an infinite amount of time has passed. (Only Penrose could ask such a question.) He speculates that after some point all the elementary particles with mass, including protons, quarks, and electrons, would decay and only photons and other massless particles would be left. If so, there would be nothing to detect the infinite passage of eternity, because photons, since they travel at the speed of light, don't experience time at all. To a photon, the eternity of the very late universe would be indistinguishable from the very early universe. The only difference would be the temperature. Admittedly, the temperature difference is enormous, but this is just a single scale. Penrose argues that a single scale doesn't matter. In a gas of photons described relationally, all that matters is the comparisons, or ratios, between things that exist at the time; the overall scale cannot be detected. So the late universe, filled with a gas of cold photons and other massless particles, becomes indistinguishable from the hot gas of the same particles filling the early universe. According to the principle of the identity of the indiscernibles, the late universe is also the birth of another universe.

This scenario of Penrose's unfolds only after an infinite time and so does not resolve the Boltzmann brain paradox. But it does predict that there would be fossils of the past universe in remnants of the Big Bang, from which we could glean information about it. While much information is wiped out by the eternity spent in thermal equilibrium, one carrier of information that never is disordered is gravitational radiation. The information carried by gravitational waves also makes it across the bounce in the cyclic models and into the new universe.

The loudest signals carried by gravitational waves are images of collisions between the great black holes that once lurked in the centers of the long-gone galaxies. These ripple outward, making great circles in the sky. They travel forever and survive the transition to the new universe. Consequently, Penrose predicts, these great circles should be visible in the cosmic microwave background, whose structure was

locked in early in our universe. These are shadows of events in the former universe.

Moreover, Penrose predicts that there should be lots of concentric circles. These come from clusters of galaxies in which, over time,
more than one pair of galactic black holes collided. This is a strik-
ing prediction, quite different from the kinds of patterns predicted by
most cosmological scenarios for the CMB. If something this unlikely
is confirmed, it would have to count as evidence for the scenario that
produced the prediction.

As of this writing, there is a controversy over whether Penrose's
concentric circles can be seen in the CMB.[8] However this turns out,
we see, once again, that cosmological scenarios in which our universe
evolved from a pre–Big Bang universe make predictions that can be
verified or falsified by observation. This is in contrast to scenarios in
which the universe is one of a simultaneous plurality of worlds — scenarios that do not, and most likely cannot, make any real predictions.

In chapter 10, I argued that a rational explanation of why particular
laws and initial conditions obtained in our universe required the selection to have happened more than once, because otherwise we could
not know why the choice was made as it was — whereas if the same initial conditions and laws occurred many times, there could be reasons
for that. I considered two ways in which the many Big Bangs could be
arranged — simultaneously or sequentially — and I argued that only in
the latter case could we expect to develop a cosmology that could answer the *Why these laws?* question while remaining scientific, in the
sense of yielding falsifiable predictions. In this chapter, I have returned
to contrast the two alternatives, and we have seen in detail that only
in the case of sequential universes are there real predictions for doable
experiments.

Thus we see that cosmology becomes more scientific, and our ideas
more vulnerable to test, when we work in a framework in which time
is real and fundamental and the history of the universe is a necessary
part of understanding its present state. Those burdened by the metaphysical presupposition that the purpose of science is to discover time-

less truths represented by timeless mathematical objects might think that eliminating time, and so making the universe akin to a mathematical object, is a route to a scientific cosmology. But it turns out to be the opposite. As Charles Sanders Peirce understood more than a century ago, *laws must evolve to be explained.*

19

The Future of Time

I N PART II, WE HAVE climbed back up from timelessness to establish time in its rightful place at the core of our conception of the world. The arguments presented in Part I for the unreality of time appeared strong, but they all depend on extending the Newtonian paradigm to a complete theory of the universe as a whole. As we have seen, the very features that make that paradigm a successful method for describing the physics of small parts of the universe undermine its application to the universe as a whole. To make further progress in cosmology (and in fundamental physics as well), we need a new conception of a law of nature, valid on the cosmological scale, which avoids the fallacies, dilemmas, and paradoxes and answers the questions that the old framework cannot address. Moreover, it must be a scientific theory—that is, it must make falsifiable predictions for new, but doable, experiments.

In chapter 10, I began the search for such a new framework by putting forward basic principles to guide our search. Paramount among them is Leibniz's principle of sufficient reason, which compels us to seek a rational reason for every choice the universe has made to be one way rather than another. This implies further principles: of the identity of the indiscernibles, explanatory closure, and no unreciprocated action. These principles frame a thoroughgoing relational approach to all the properties of things in nature.

I then argued that the only way to realize these principles and discover a workable cosmological theory is to hypothesize that the laws of nature evolve over time. This requires that time be real, and global. One promising development is shape dynamics which, as described in chapter 14, evokes a preferred global notion of time from within general relativity.

The notion of a real time, within which laws of nature evolve, together with our principles, gives us a foundation for a new cosmological theory. The developments described in Part II in chapters 11 through 18 are not yet fact and do not yet amount to a coherent theory. They are instead a vision of how we might reconceive both the universe and the task of cosmology. Each is speculative, but several make genuinely testable predictions for doable experiments. Whether any are confirmed by experiment or not, they at least demonstrate that the hypothesis of the reality of time leads to a more scientific cosmology.

The notion of a real and global time is also helpful in resolving other unsolved problems in physics. For example, we need to go beyond the statistical prediction of quantum mechanics to describe and explain what happens in individual events. In chapters 12 and 13, I described two new approaches to a deeper theory of quantum phenomena, both of which require time to be fundamental. These approaches appear to differ from quantum mechanics sufficiently that they could be distinguished from it experimentally.

Another arena where real time operates is in the description of behavior in the macro world, where thermodynamics emerges along with such concepts as temperature, pressure, density, and entropy. At this non-quantum level, time appears to be strongly directional, and we can discern several arrows of time that distinguish the past strongly from the future. In a theory where time is inessential or emergent, this fact of the universe's time-asymmetry is baffling. It forces us to attribute the most obvious and apparent features of the world to an extremely improbable choice of initial conditions. The difficulty can be avoided by presuming that time is real and that the fundamental theory is as asymmetric in time as the universe reveals itself to be.

However, it's one thing to say that time is real but another to say that it makes sense to talk about what's happening "right now" throughout the universe — that is, simultaneously with our experience of time passing. The idea of a global time means that our experience of time passing is shared across the universe, but of course it conflicts directly with the relativity of simultaneity of special and general relativity. This conflict must be faced, because the relativity of simultaneity, along with the idea that reality is a shared notion, leads, as we saw in chapter 6, to the block-universe picture, within which the most basic aspect of our experience — the passage of time — is not real.

One might try to imagine a sense in which time is real that doesn't conflict with the relativity of simultaneity — but it would require either a solipsistic or an observer-dependent notion of reality, in which the distinction between the real present and the yet-to-be-real future is not an objective property shared by all observers. And, as I have emphasized, the global-time hypothesis helps a great deal in going beyond quantum theory and understanding space as emergent. It's also important to note that the global-time hypothesis need not conflict with the experimental confirmations of special relativity, as we saw was true in shape dynamics. In the end, the hypothesis that there is a preferred global time in nature is one to be settled by experiment, which is why I have espoused hypotheses that can lead to new predictions by which they can be checked.

◆

The idea that laws evolve has the promise of making fundamental physics more predictive. But it brings with it one final dilemma. It is natural to ask whether there is a law that governs how the laws evolve. We can call such a law, which acts on laws rather than directly on elementary particles, a *meta-law*. It might be hard to observe the action of this meta-law, as it may act only during violent episodes such as the

Big Bang. However, if we want a complete explanation for our universe, one that fully realizes the ambition of the principle of sufficient reason, shouldn't there be such a meta-law?

But suppose there is a meta-law. Shouldn't we want to know why this meta-law, rather than a different one, governs the evolution of laws in our universe? And if a meta-law may act on past laws to produce laws in the future, part of the explanation for what the laws are presently will depend on what those past laws were, so we can't avoid the *Why these initial conditions?* question. The meta-law hypothesis could lead to an infinite regression (*Why this meta-law?* might be answered by meta-meta-laws, and so on). This is one horn of the dilemma. The other is the possibility that there *is* no meta-law. There would then be an element of randomness in the evolution of laws, the result again being that not everything is explainable and the principle of sufficient reason is flouted at the very foundations of science. Roberto Mangabeira Unger and I call this the *meta-laws dilemma.*

It might look at first like a dead end, but after living with it for several years I have come to believe that it is, instead, a great scientific opportunity, a provocation to invent a new kind of theory that will resolve it. I'm convinced that the meta-laws dilemma is solvable and that how it is solved will be the key to the breakthroughs that will enable cosmology and fundamental physics to progress in this century.

The meta-laws dilemma is temporarily circumvented by cosmological natural selection (see chapter 11) when a limited and statistical meta-law is hypothesized. When I postulated that the parameters of the Standard Model change by small random amounts at each bounce, I described a kind of meta-law that partly circumvents the dilemma. Certainly we want to know more about how this happens and be able to describe the mechanism generating the random parameter changes. More insight into this might be provided by a quantum theory of gravity, such as loop quantum gravity or string theory (the latter being the context within which this idea was first conceived). But even without further insight, the hypothesis of cosmological natural selection is both explanatory and falsifiable.

The principle of precedence is another approach to a meta-law.

By being partly statistical, it, too, circumvents — or at least post-pones — the meta-laws dilemma. Even the postponement of a dilemma can be fruitful, opening a space for hypotheses that can be investigated experimentally and, in turn, suggest new questions and approaches. But to ultimately solve the meta-laws dilemma, the dynamics by which laws evolve must be sufficiently different from the laws we're familiar with so that the *Why this meta-law?* and *Why these initial conditions?* questions do not arise.

Here's one approach that solves the dilemma in a surprising way: Suppose that any two proposals for a meta-law would be equivalent to each other — that is, have identical effects on how the laws evolve.[1] There might be a *principle of the universality of meta-law,* just as there's universality in computation. In that sphere, "universality" means that any function that can be computed by one computer can be computed by any other computer, no matter what operating system it's running. The idea of universality for meta-laws is analogous, holding that there is no meaning to which meta-law is operating, since all the experimental predictions will be the same regardless.

Still another approach to a science of cosmology that goes beyond the Newtonian paradigm is to imagine a marriage of law and configuration. There would not be two things to know — the law and the state — but only one, which unifies them into a *meta-configuration* that contains information about both. This idea accords with the hypothesis that all that's real is real in the present moment. To the extent that a law is acting, its specification is part of the present moment. The specification of a law and of a configuration cannot be too different, so we unify them into a single meta-configuration. Just as Galileo unified the heavenly and earthly domains, it may be time to unify their shadow, which is the distinction between timeless law and timebound configuration.

Evolution of the meta-configuration would be driven by a rule so simple that it is explained by a principle of universality. The choice of initial configuration would specify the initial law as well as the initial conditions. There would be aspects of the configuration that evolve quickly and aspects that evolve much more slowly. The former aspects would count as the configuration, which would evolve via what we might call laws, specified by the slower-moving aspects. But over longer time scales, the distinction between laws and configurations would break down. I have developed a simple model of this idea that is not, so far, very realistic.[2]

These two ideas, together with the principle of precedence and cosmological natural selection, already give us four ways to address the meta-laws dilemma. They are admittedly first steps. It is not an exaggeration to say that the direction of 21st-century cosmology will be determined by how the meta-laws dilemma is resolved.

◆

In the opening chapter, I raised some questions about the role that mathematics plays in science. Before we close, I want to briefly come back to this subject, because it should be clear that the reality of time has important implications for the role of mathematics in physics.

Within the Newtonian paradigm, a timeless configuration space can be described as a mathematical object. The laws can also be represented by mathematical objects, as can their solutions, which are possible histories of the system. What the mathematics corresponds to are not the actual physical processes but only records of them once completed — which are also, by definition, timeless. Yet the world remains, always, a bundle of processes evolving in time, and only small parts of it are representable by timeless mathematical objects.

Because the Newtonian paradigm cannot be scaled up to include the universe as a whole, there need be no mathematical object corresponding to the exact history of the entire universe. Nor, for the universe as a whole, need there be a timeless configuration space and timeless laws represented as timeless universal mathematical objects.

John Archibald Wheeler used to write physics equations on the blackboard, stand back, and say, "Now I'll clap my hands and a universe will spring into existence." Of course, it didn't.[3] Stephen Hawking asked, in *A Brief History of Time*, "What is it that breathes fire into the equations and makes a universe for them to describe?" Such utterances reveal the absurdity of the view that mathematics is prior to nature. Math in reality comes after nature. It has no generative power. Another way to say this is that in mathematics conclusions are forced by logical implication, whereas in nature events are generated by causal processes acting in time. This is not the same thing; logical implications can model aspects of causal processes, but they're not identical to causal processes. Logic is not the mirror of causality.

Logic and mathematics capture aspects of nature, but never the whole of nature. There are aspects of the real universe that will never be representable in mathematics. One of them is that in the real world it is always some particular moment.

So one of the most important lessons that follow once we grasp the reality of time is that nature cannot be captured in any single logical or mathematical system. The universe simply is — or better yet, happens. It is unique. It happens once, as does each event — each unique event — that nature comprises. Why it is, why there is something rather than nothing, is probably not a question that has an answer — save that, perhaps, to exist is to be in relation to other things that exist and the universe is simply the set of all those relations. The universe itself has no relation to anything outside it. The question of why it exists rather than not is beyond the scope of the principle of sufficient reason.

What form are the discoveries of cosmology to be expressed in, if not in a single timeless mathematical law acting on a timeless space of initial conditions? This is a question on which the future of cosmology depends. With a little thought, some possible answers emerge.

The examples I have given, such as cosmological natural selection and the principle of precedence, demonstrate that we can conceive of testable scientific theories that go beyond the Newtonian paradigm. It is good to reflect on the fact that in the history of science there are many hypotheses that don't need to be stated mathematically. And in some cases mathematics is not needed to work out their consequences. An example is the theory of natural selection; aspects of it have been captured in simple mathematical models, but no single model captures the whole variety of mechanisms by which natural selection acts in nature. Indeed, new mechanisms of evolution may emerge at any time, as new species are born.

To be scientific, hypotheses must suggest observations by which they could be verified or falsified. Sometimes this requires expression in mathematics; sometimes it doesn't. Mathematics is one language of science, and it is a powerful and important method. But its application to science is based on an identification between results of mathematical calculations and experimental results, and since the experiments take place outside mathematics, in the real world, the link between the two must be stated in ordinary language. Mathematics is a great tool, but the ultimate governing language of science is language.

◆

The challenge facing us should not be underestimated. Cosmological science is in a crisis, and the only sure bet is that proceeding on the basis of the methodologies that have served us so well until now will not get us anywhere. We can see from the paradoxes that ensue what happens if we try to take the standard Newtonian paradigm as the basis of cosmology. So we have to go forward into the unknown. We face a choice among radical programs. Which one turns out to be correct can only be decided once we see which direction leads to testable predictions for new observations and those observations are made. We will also expect any new theory to provide robust explanations of known, if presently mysterious, facts. We should encourage diverse approaches to these hard questions.

But the choices are nonetheless stark. To contrast the choices before us, on the next two pages I list pairs of contrary assertions that we have encountered in this book. These frame the implications of taking time as an illusion or as the core of reality.

Time is an illusion. Truth and reality are timeless.

Space and geometry are real.

Laws of nature are timeless and inexplicable, apart from selection by the anthropic principle.

The future is determined by the laws of physics acting on the initial conditions of the universe.

The history of the universe is, in all its aspects, identical to some mathematical object.

The universe is spatially infinite. Probabilistic predictions are problematic, because they come down to taking the ratio of two infinite quantities.

The initial singularity is the beginning of time (when time is defined at all) and is inexplicable.

Our observable universe is one of an infinite collection of simultaneously existing but unobservable universes.

Equilibrium is the natural state and inevitable fate of the universe.

The observed complexity and order of the universe is a random accident due to a rare statistical fluctuation.

Quantum mechanics is the final theory and the right interpretation is that there are an infinity of actually existing alternative histories.

Nothing in science is certain. What we can do, however, in the face of uncertainty is try to construct reasoned arguments for diverse hypotheses. This is what I have done here. And while the ultimate test is experiment, we can draw some conclusions from how generative a research program is of new hypotheses and of predictions by means of which they can be checked.

Time is the most real aspect of our perception of the world. Everything that is true and real is such in a moment that is one of a succession of moments.

Space is emergent and approximate.

Laws of nature evolve in time and may be explained by their history.

The future is not totally predictable, hence partly open.

Many regularities in nature can be modeled by mathematical theories. But not every property of nature has a mirror in mathematics.

The universe is spatially finite. Probabilities are ordinary relative frequencies.

The Big Bang is actually a bounce which is to be explained by the history of the universe before it.

Our universe is a stage in a succession of eras of the universe. Fossils, or remnants, of previous eras may be observed in cosmological data.

Only small subsystems of our universe come to uniform equilibria; gravitationally bound systems evolve to heterogeneous structured configurations.

The universe naturally self-organizes to increasing levels of complexity, driven by gravitation.

Quantum mechanics is an approximation of an unknown cosmological theory.

The research program based on the timeless universe that embraces quantum mechanics and the multiverse as the final theory has been around for more than two decades. It has not yet produced a single falsifiable prediction for a currently doable experiment. At best, it has produced speculations about a novel phenomenon, collisions of bubble universes, remnants of which might be observed if we're lucky.

However, these speculations are not falsifiable predictions, because failure to verify the predictions can be easily explained away at no cost to the speculation. Nor have the basic difficulties this program faces been resolved, despite many years of work by smart and determined scientists. These difficulties have to do with making predictions when the universe is one of infinitely many universes, all but one of them unobservable; with the definition of probabilities when there are an infinite number of copies of every event; and with the basic fact that neither theory nor observation much constrains the invention of scenarios about things that might be true beyond the range of our observations.

It's impossible to be sure that nothing important will come from the investigations of these ideas, but it seems likely that history will describe them as failures — failures due to a misconceived approach to a foundational problem in science. The failure arises from taking a method suitable for studying small parts of the universe and applying it to the whole of existence.

If I have characterized it correctly, the failure is not superficial and cannot be fixed just by inventing another scenario of the same kind. Cosmological questions such as *Why these laws?* and *Why these initial conditions?* cannot be answered by a method that takes the laws and initial conditions as input. The remedy must be radical, involving not just the invention of a new theory but a new method and hence a new kind of theory.

While the task is daunting, we do have several things on our side. The first, and most likely primitive, attempts to frame hypotheses about the evolution of laws — hypotheses that involve a possible history of the universe before the Big Bang — have led to falsifiable predictions for doable observations. These include the predictions of cosmological natural selection and the predictions of cyclic cosmologies. It's too early to tell whether any of these ideas are true, but it's encouraging to know that present and near-future observations could lead us to reject them as false. These simple examples suggest that scenarios in which the universe is a stage in a succession of universes are testable and hence scientific.

The other thing we have going for us is the wisdom of the deepest cosmological thinkers in history, particularly Leibniz, Mach, and Einstein. From them, we get several principles that have served so far as excellent guides to the development of physics.

The most radical suggestion arising from this direction of thought is the insistence on the reality of the present moment and, beyond that, the principle that all that is real is so in a present moment. To the extent that this is a fruitful idea, physics can no longer be understood as the search for a precisely identical mathematical double of the universe. That dream must be seen now as a metaphysical fantasy that may have inspired generations of theorists but is now blocking the path to further progress. Mathematics will continue to be a handmaiden to science, but she can no longer be the Queen.

The reward we get for sacrificing a queen is a more democratic vision of the furniture of physical theories. Just as the distinction between royalty and commoners was discarded long ago, so must we reject and go beyond an absolute distinction between the states of affairs in the world and the laws by which they evolve in time. No longer can absolute, timeless laws be seen to dictate the evolution of the time-bound configuration of the world. If everything that is real is real in a moment, then the distinction between laws and states must be a relative one, which arises and is discernible in relatively cold and calm cosmological eras like our own. But in other, more violent eras, the distinction must dissolve into a new, fully dynamic description of the world which is rational and answers to the principle of sufficient reason.

By allowing laws to evolve in time, we increase our chances of explaining them by hypotheses that have testable consequences. It might seem that having laws evolve weakens their power, but in fact it increases the overall power of science, whereas extending ideas that work in the Newtonian paradigm to the domain of cosmology weakens the power of science. If we admit into our conception of nature evolution and time at the deepest levels, we are more apt to comprehend this mysterious universe in which we find ourselves.

Will this new path succeed? Only time will tell.

EPILOGUE

Thinking in Time

ALL THE PROGRESS of human civilization, from the invention of the first tools to our nascent quantum technologies, is the result of the disciplined application of the imagination. Imagination is the organ that allows us to thrive on the cusp between danger and opportunity; it is an adaptation to the reality of time. We are superb hunters and gatherers and processors of information, but we are far more than that: We have a capacity for imagining situations that are not implied by the data we have. Our imagination lets us anticipate dangers before they're imminent, which means we can plan to meet them. We're no match for the tiger in the night, and there's nothing we can do to keep her from making a meal of our children once she's pounced. But because we imagined it, we built a fire that kept her away.

To know that we can build a fire to keep the tigers away may not seem impressive, but think back a few hundred thousand years to the person who first did it. At the time, it must have seemed insane to use one deadly menace to keep another away. Just the idea that fire could be controlled must have taken tremendous imagination and courage. In the modern world, we live with fire hidden throughout our homes, in the wires in the walls, in the stove, in the furnace in the basement. We don't even think of it — or at least not until we're in the car on our way somewhere and wondering whether we turned the oven off. But if we hadn't descended from people who, hundreds of thousands of years ago, imagined ways to harness fire, we would still be prey.

This is the grand bargain of human life: to thrive on the cusp of uncertainty. We thrive on the boundary between opportunity and danger and live with the knowledge that we can't control everything or keep bad stuff from happening every now and then.

The other animals evolved to be in sync with their environments. For them, surprise is nearly always bad news, for it signals a change in the environment that exposes them to danger for which they're not adapted. At some point in our evolution, our ancestors evolved the organ of imagination, which let us adapt to novel environments. Imagination enabled us to turn change and surprise into opportunities to extend our domain across the planet.

Some 12,000 years ago, we adapted our environments to ourselves, becoming farmers rather than opportunistic hunter-gatherers. Since then, our footprint has extended to the point where our impositions on the Earth's natural systems threaten to cause us great harm. Because imagination is our game, and imagination got us here, only imagination can provide the new ideas that will take us safely through the surprises to come.

The same imagination that drove our adaptation led to the essentially tragic aspect of human life, which is that we can imagine our own inevitable death. Wanting, needing, to survive as long as possible, we push back against the inevitable, and because we're human we overshoot, and not just by a little. One result is the flowering of civilizations, science, the arts, the wonderful technologies we take almost for granted. Another is all the waste that our overshooting produces, because the most reliable safeguard against exponential decline is exponential growth. So a species that evolved to fit a relatively narrow and rare niche has conquered the entire surface of the planet. Our closest relatives are nearly endangered species living in a few forests in Africa, but there are billions of us. The speciation that divided us from the other primates is often attributed to "culture," but isn't that just another word for our incessant imagining and striving for better ways to live?

We can envision beings who don't ask for much, who take minimally from their environments and societies, who instinctively live in balance with their worlds. Some of us wish to become like that, and to live more simply is indeed good advice, but on the whole that's not the way of human beings. Our way is to aspire always to more than and other than what we have. To be human is to imagine what is not, to seek beyond the limits, to test the constraints, to explore and rush and tumble across the intimidating boundaries of our known world.

There's a romantic idea that the breaching of boundaries and living out of balance with our environment is a pathology of capitalism and modern technological society. But it's just not so. As Stone Age conquerors of North America, we surged across the continent, wiping out most of the large mammals as we went. A much larger proportion of hunter-gatherers died in tribal warfare than the proportion of Europeans slain in the two world wars of the 20th century.

As a species, we seem to be at the peak of our dominance of the planet's ecosystems and resources. We all know that the present situation is unsustainable. Unsustainability was bound to happen; it is always the result of exponential growth. We're just the fortunate ones who live within a lifetime of the peak and the crisis that will follow if we don't learn fast to act more wisely than in the past. If we persist in thinking outside time, we will not surmount the unprecedented problems raised by climate change. We cannot rely on the standard menu of political solutions, because those problems are defined by the failure of our present political systems. Only by thinking in time do we have a chance to thrive for centuries more.

There was someone who for the first time had the courage to make her children safe by harnessing fire. Who will have the courage to realize that the safety of our children may depend on our learning to steer the climate?

◆

Let's imagine it's 2080 and the problems of climate change have been faced and ameliorated. Our children will be elderly — or perhaps, due to medical advances, still in the prime of life. How will their thinking have changed because of their avoidance of catastrophe?

It's easier to imagine what their perspective will be if we do nothing to bring carbon-dioxide emissions under control. As they face rising temperatures and sea levels, drought and failing crops, as the northern cities crowd with refugees, you can well enough imagine what they'll wish they could say to us.

But suppose we've found the wisdom to avoid all this. What will we have learned along the way that made success possible? And what positive good (as opposed to the avoidance of disaster) will society have achieved by solving this crisis? The literature on climate change is typically couched in negatives. Over and over, we read about the dreadful consequences of inaction, but nowhere can we find a discussion of the ancillary benefits that accrue when we solve this problem. People who exercise and eat well find positive effects in being healthy that outweigh the motives of avoiding illness and early death. Might there likewise be positive benefits of living within an economy that fosters a healthy planet?

The consequences of overcoming the climate crisis are difficult to predict, because to succeed we have to do more than solve a global engineering problem. Even among those who appreciate the seriousness of the crisis, adherence to one or another of two opposing viewpoints, both false, delays real progress. For those who see the world in economic terms, nature is a resource to be exploited and transcended — and climate change is just an agricultural problem on a larger scale, to be managed by cost-benefit analysis. For environmental activists, nature is paramount and pristine, and can only be diminished by the encroachments of civilization; climate change, for them, is just another issue of preservation. Both miss the point, because both assume that nature and technology are mutually exclusive categories, so that when they clash a choice must be made between them. But an adequate solution to the crisis requires muddying the distinction between the natural and the artificial. It requires not a choice between nature and technology, but a reorientation of their relationship to each other.

The overwhelming scientific consensus tells us that it is we who are now destabilizing the climate, but it is also true that the climate has fluctuated suddenly in the past between very different states. If this happens again — whether triggered by our doings or not — it will have dire consequences for us. Because we're able to prevent or moderate major changes in the climate, we must do so — for the same reason that we must look out for and destroy asteroids that might collide with Earth. After we have resolved this emergency, we will be committed to a continuing regulation of the climate, to keep it in a range within which humanity can thrive. This means melding our technologies with the natural cycles and systems that already regulate the climate.

Once we understand how the natural systems regulating the climate react to our technologies, and we begin to operate our technologies and economies so that they work in harmony with the climate, we will have transcended the divide between the natural and the artificial on a planetary scale. The economy and the climate will be aspects of a single system. To survive the climate crisis, we have to conceive of and establish a new kind of system, a symbiosis of the natural processes determining the climate with our technological civilization.

We're accustomed to seeing ourselves as apart from nature and our technologies as impositions on the natural world. But whether we fantasize about our conquering nature or nature surviving us, we have reached the limits of the usefulness of the idea that we're separate from nature. If we want to survive as a species, we need a new way of seeing ourselves, in which we and everything we make and do are as natural as the cycles of carbon and oxygen we emerged from and in which we participate with every breath.

To begin this task, we have to understand the roots of the distinction between the artificial and the natural. These have a great deal to do with *time.* The false idea we have to put behind us is the idea that what is bound in time is an illusion and what is timeless is real.

Early expressions of this perennial philosophy are found in Christian interpretations of Aristotelian and Ptolemaic cosmology, in which, as I described in chapter 1, the earthly sphere is the unique abode of life but also of death and decay, surrounded by perfect spheres of unchanging crystalline construction that rotate eternally around the

Earth, carrying the moon, the sun, and the planets. The stars are fixed to the outermost sphere, above which live God and His angels. From this scenario springs the prevailing notion that goodness and truth are to be found above us, evil and falseness below. To learn to live with our planet, we have to rid ourselves of the vestiges of this old yearning for elevation from it.

The same hierarchy applies to the natural/artificial divide, although different people see it differently. Some value the artificial over the natural world of living things because — being the product of minds rather than of mindless, messy evolution — it is closer to absolute perfection and therefore closer to timelessness. Others esteem the natural as having a purity lacking in artificial constructions.

How can we get rid of the conceptual structure of a divided and hierarchical world separating the natural and artificial? To escape this conceptual trap, we need to eliminate the idea that anything is, or can be, timeless. We need to see everything in nature, including ourselves and our technologies, as time-bound and part of a larger, ever evolving system. A world without time is a world with a fixed set of possibilities that cannot be transcended. If, on the other hand, time is real and everything is subject to it, then there is no fixed set of possibilities and no obstacle to the invention of genuinely novel ideas and solutions to problems. So to move beyond the distinction between the natural and artificial and to establish systems that are both, we have to situate ourselves in time.

We need a new philosophy, one that anticipates the merging of the natural and the artificial by achieving a consilience of the natural and social sciences, in which human agency has a rightful place in nature. This is not relativism, in which anything we want to be true can be. To survive the challenge of climate change, it matters a great deal what is true. We must also reject both the modernist notion that truth and beauty are determined by formal criteria and the postmodern rebellion from that, according to which reality and ethics are mere social constructions. What is needed is a relationalism, according to which the future is restricted by, but not determined by, the present, so that novelty and invention are possible. This will replace the false hope of transcendence to a timeless, absolute perfection with a genuinely hopeful view of an ever expanding realm for human agency, within a cosmos with an open future.

Part of the program of the new philosophy is to save cosmology from an unscientific excursion by recognizing the central role time plays on a cosmological scale. That scientific task has been the focus of this book. But equally important, a civilization whose scientists and philosophers teach that time is an illusion and the future is fixed is unlikely to summon the imaginative power to invent the communion of political organizations, technology, and natural processes — a communion essential if we are to thrive sustainably beyond this century.

◈

Probably the greatest harm done by the metaphysical view that reality is timeless is through its influence on economics.[1] The basic flaw in the thinking of many economists is that a market is a system with a single equilibrium state. This is a state where the prices have adjusted so that the supply of each good exactly meets the demand for it, according to the law of supply and demand. Further, such a state is said to optimize everyone's satisfaction. There's even a mathematical theorem holding that in equilibrium no one can be made happier without making someone else less happy.[2]

If each market has one and only one such equilibrium point, then the wise and ethical thing to do is leave the market alone so it can settle at that point. Market forces (i.e., the way producers and consumers respond to changes in prices) should be sufficient to do this. A recent version of this idea is the *efficient-market hypothesis,* which holds that prices reflect all the information relevant to the market. In a market with many players contributing their knowledge and views by means of their bids and asks, it is impossible that any asset is mispriced for long. Remarkably, this line of reasoning is supported by elegant mathematical models within which are formal proofs that equilibrium points always exist; that is, there are always choices of prices such that supply exactly balances demand.

This simple picture, in which the market always acts to restore conditions to equilibrium, depends on the assumption that there's only one equilibrium. But this isn't the case. Economists have known since the 1970s that their mathematical models of markets have typically many equilibrium points where supply balances demand. How many? The number is hard to estimate but certainly grows at least proportionally to the numbers of companies and consumers, if not faster. In a complex modern economy, with many goods made by many firms and bought by many consumers, there are a lot of ways to set the prices of goods so that supply and demand are in balance.[3]

Because there are many equilibria where market forces balance, they cannot all be completely stable. The question is, then, how a society chooses which equilibrium to be in. The choice cannot be explained solely by market forces, because supply and demand are balanced in each of the many possible equilibria. Regulations, laws, culture, ethics, and politics then play a necessary role in determining the evolution of a market economy.

How is it possible that influential economists have argued for decades from the premise of a single, unique equilibrium, when results in their own literature by prominent colleagues showed this to be incorrect? I believe the reason is the pull of the timeless over the timebound. For if there is only a single stable equilibrium, the dynamics by which the market evolves over time is not of much interest. Whatever happens, the market will find the equilibrium, and if the market is perturbed, it will oscillate around that equilibrium and settle back down into it. You don't need to know anything else.

If there is a unique and stable equilibrium, there's not much scope for human agency (apart from each firm maximizing its profits and each consumer maximizing his pleasure) and the best thing to do is to leave the market alone to achieve that equilibrium. But if there are many possible equilibria, and none is completely stable, then human agency has to participate in and steer the dynamics by which one equilibrium is chosen out of many possibilities. In the thinking of the economic gurus who won the day for deregulation, the role of human agency was neglected, in deference to an imagined mythical timeless state of nature. This was the profound conceptual mistake that opened the way for the errors of policy that led to the recent economic crisis and recession.

Another way to speak about this mistake is in terms of path dependence and path independence. A system is path-*dependent* if it matters how the system evolved from one configuration to another — that is, our present circumstances depend not just on where we are but on how we got here. A system is path-*independent* if everything depends only on its current configuration and nothing depends on how it got here. In a path-independent system, time and dynamics play little role, because at any time the system is either in its unique state or fluctuating slightly around it. In a path-dependent system, time plays an important role.

Neoclassical economics conceptualizes economics as path-independent. An efficient market is path-independent, as is a market with a single, stable equilibrium. In a path-independent system, it should be impossible to make money purely by trading, without producing anything of value. This sort of activity is called arbitrage, and basic financial theory holds that in an efficient market arbitrage is impossible, because everything is already priced in such a way that there are no inconsistencies. You cannot trade dollars for yen, trade those for euros, and trade euros back for dollars and make a profit. Nonetheless, hedge funds and investment banks have made fortunes trading in currency markets. Their success should be impossible in an efficient market, but this does not seem to have bothered economic theorists.

Decades ago, the economist Brian Arthur, at the time the youngest holder of an endowed chair at Stanford University, began to argue that economics was path-dependent.[4] His evidence for this was that the economic dictum known as the *law of decreasing returns* is not always correct. This law says that the more of something you produce, the less profit you make from each item you sell. This is not necessarily true, for example, in the software business, where it costs almost nothing to make and distribute additional copies of a program, so all the costs are up front. Arthur's work was treated as heresy — and indeed, without the assumption of decreasing returns, some of the mathematical proofs in neoclassical economic models fall apart.

In the mid-1990s, an economics graduate student at Harvard, Pia Malaney, working with the mathematician Eric Weinstein, found a mathematical representation of the path-dependence of economics. In geometry and physics, there's a well-understood technique for studying path-dependent systems which is called *gauge fields;* these provide the mathematical foundations for our understanding of all the forces in nature. Malaney and Weinstein applied this method to economics and found that it is path-dependent. Indeed, there's an easily computable quantity called the *curvature* that measures path-dependence, and they found it was not zero in typical models of markets where prices and consumer preferences change. Hence, like the Earth and the geometry of spacetime, the mathematical spaces that model markets are curved. In her PhD thesis, Malaney applied their model to increases in the Consumer Price Index and found it had been miscomputed by economists who did not take path dependence into account in their economic models.[5]

The work of Malaney and Weinstein was ignored by academic economists, but the path dependence of markets has since been rediscovered by a number of physicists who found it natural to apply gauge theories to them.[6] There's no way to know how many hedge funds are making money discovering arbitrage opportunities by measuring curvature — that is, path dependence that's not supposed to exist in neoclassical economics — but this is doubtless going on.

A path-dependent market is one where time really matters. How does neoclassical economic theory deal with the fact that in reality markets do evolve in time, in response to changing technologies and preferences, continually opening up opportunities to make money that are not supposed to exist in their models? Neoclassical economics treats time by abstracting it away. In a neoclassical model, you as a consumer are modeled by a utility function. This is a mathematical function that gives a number to every possible combination of goods and services that might be purchased in the economy you live in. This is a huge set, but, hey, this is math, so let's continue. The idea is that the more utility a collection of goods and services has for you, the more of them you'll want to buy. The models then assume that you buy the collection of goods and services that maximizes your desires as measured by your utility function, given the constraint of how much you can afford.

What about time? The idea is that the lists of goods and services includes all the goods and services you might want to buy over your entire lifetime. So the budget constraint imposed is over your lifetime income. Now, this is clearly absurd; how can people know what they'll want or need decades from now, or what their lifetime income will be? The models deal with such contingencies — the fact that over a lifetime one confronts a myriad of unpredictable circumstances — by lumping them into the lists of goods and services. That is, they assume there's a definite price for every possible collection of goods and services at every time and in every situation that might arise — even decades into the future. There is a price, say, not just for a Ford Mustang but for a Ford Mustang in 2020 under every possible contingency. The models also assume not just that all the goods and services we might buy now have been perfectly priced in equilibrium but that every future price of any collection of goods and services under every possible contingency is also perfectly priced. Moreover, they assume that there are so many investors with such a variety of views that the bets cover the whole space of possible contingencies and positions — whereas investigations of real markets have found that a small number of positions are occupied by most of the traders.[7]

That the neoclassical economic models go to such absurd lengths to abstract time and contingency only shows how central the issue of time is. There is a powerful, if unacknowledged, attraction to theories in which time plays no role — perhaps because they give the theorists a sense of inhabiting a timeless realm of pure truth, against which the time and contingency of the real world pale.

We live in a world in which it is impossible to anticipate most of the contingencies that will arise. Neither the political context, nor the inventions, nor the fashions, nor the weather, nor the climate are precisely specifiable in advance. There is, in the real world, no possibility of working with an abstract space of all the contingencies that may evolve. To do real economics, without mythological elements, we need a theoretical framework in which time is real and the future is not specifiable in advance, even in principle. It is only in such a theoretical context that the full scope of our power to construct our future can make sense.

Furthermore, to meld an economy and an ecology, we need to conceive of them in common terms — as open complex systems evolving in time, with path dependence and many equilibria, governed by feedback. This fits the description briefly given here of economics, and it fits the theoretical framework of ecology as well, with the climate as the sum and expression of a network of chemical reactions driving and regulated by the basic cycles of the biosphere.[8]

◈

One of the difficulties we face when we attempt to have a constructive conversation about the future is that our present culture is characterized by incoherence. People on one frontier of knowledge hardly know what seekers on the other frontiers are talking about. Our conversations are siloed. Most physical scientists don't know much about the breakthroughs in biology, let alone what's happening on the forefront of social theory, and don't have a clue about what questions influential artists ask each other.

If our civilization is to thrive, it would be helpful to base our decision making on a coherent view of the world, in which, to begin with, there is consilience between the natural and social sciences. The reality of time can be the foundation for this new consilience, in which the future is open and novelty is possible on every scale from the fundamental laws of physics to the organization of economies and ecologies.

In the past, great conceptual steps in physical science have been echoed in social science. Newton's idea of absolute time and space is said to have greatly influenced the political theory of his contemporary, John Locke. The notion that the positions of particles were defined with respect not to each other but to absolute space was mirrored in the notion of rights defined for each citizen with respect to an unchanging absolute background of the principles of justice.

General relativity moved physics to a relational theory of space and time, in which all properties are defined in terms of relationships. Is this mirrored in an analogous movement in social theory? I believe that it is and that it can be found in the writings of Unger and a number of other social theorists. These explore, in the context of social theory, the implications of a relational philosophy according to which all properties ascribed to agents in a social system arise from their relationships and interactions with one another. As in a Leibnizian cosmology, there are no external timeless categories or laws. The future is open, because there is no end to the novel modes of organization that may be invented by a society as it continually confronts unprecedented problems and opportunities.

This new social theory attempts to refashion democracy into a global form of political organization able to guide the evolution of the burgeoning multiethnic and multicultural societies. This refashioned democracy must also be up to the task of making the necessary decisions to survive the global crises posed by climate change.

Here's my understanding of what democracy looks like from the relational perspective of the new philosophy. Remarkably, the same ideas provide an understanding of how science works. This is important, because the challenge of climate change requires the interaction of science and politics.

Both democratic governance and the workings of the scientific community have evolved to manage several basic facts about human beings. We're smart, but we're flawed in characteristic ways. We're able to study our situation in nature over a single lifetime and accumulate knowledge over many lifetimes. But we have also evolved a capacity for thinking and acting at the snap of a twig. This means we often make mistakes and fool ourselves. To combat our propensity for error, we have evolved societies that embrace the contradiction between the conservative and the rebel in the service of future generations. The future is genuinely unknowable, but one thing we can be fairly sure of is that our descendants will know a lot more than we do. By working within communities and societies, we can achieve much more than we can as individuals, yet progress requires individuals to take great risks to invent and test new ideas.

Scientific communities, and the larger democratic societies from which they evolved, progress because their work is governed by two basic principles.[9]

(1) When rational argument from public evidence suffices to decide a question, it must be considered to be so decided.

(2) When rational argument from public evidence does not suffice to decide a question, the community must encourage a diverse range of viewpoints and hypotheses consistent with a good-faith attempt to develop convincing public evidence.

I call these the *principles of the open future.* They underlie a new, pluralistic stage of the Enlightenment—a stage now arising. We respect the power of reasoning when it's decisive, and when it isn't we respect those who in good faith disagree with us. The limitation to people of good faith means people within the community who accept these principles. Within such communities, knowledge can progress, and we can strive to make wise decisions about a future that is not completely knowable.

◆

Even given perfect adherence to the principles of the open future, science is unlikely ever to solve some of the questions we'd most like the answers to.

Why is there something rather than nothing? I can't imagine anything that would serve as an answer to this question, let alone an answer supported by evidence. Even religion fails here, for if the answer is "God," there was something—God, that is—to begin with. Or, *If time has no beginning, do all causes recede into the infinite past?* Is there no final reason for things? These are real questions, but if they have answers, those are likely to forever remain outside science. Then there are questions that science cannot answer now but that are so clearly meaningful that sometime in the future, it is hoped, science will evolve language, concepts, and experimental techniques to address them.

I have argued that everything that is real and true is such in a moment that is one of a succession of moments. But what is it that's real? What is the substance of these moments and the processes that connect them?

We can agree that the universe is not identical or isomorphic to a mathematical object, and I've argued that there is no copy of the universe, so there is nothing that the universe is "like." What, then, is the universe? Although any metaphor will fail us, and every mathematical model will be incomplete, nonetheless we want to know what the world consists of. Not *What is it like?* but *What is it?* What is the substance of the world? We think of matter as simple and inert, but we don't know anything about what matter really is. We know only how matter interacts. What is the essence of the existence of a rock? We don't know; it's a mystery that each discovery about atoms, nuclei, quarks, and so on only deepens.

I would dearly love to know the answer to this question. Sometimes I think about what a rock is when I'm trying to go to sleep, and I comfort myself with the idea that there must be, somewhere, an answer to what the universe is. But I have no idea how to look for it, whether through science or another route. It is so easy to make stuff up, and the bookshelves are full of metaphysical proposals. But we want real knowledge, which means there must be a way to confirm a proposed answer. This limits us to science. If there's another route to reliable knowledge of the world besides science, I'm unlikely to take it, because my life is centered around a commitment to the ethics of science.

As for science itself, we can't predict the future (that's the point of this book), but the relationalist view makes me doubt that science can tell us what the world really is. This is because relationalism claims that the quantities physics can measure and describe all concern relationships and interactions. When we ask about the *essence* of matter, or of the world, we are asking what it is intrinsically — what it is in the absence of relationships and interactions.[10] The relationalist stance is that there's nothing real in the world apart from those properties defined by relationships and interactions. Sometimes this idea seems compelling to me; at other times it seems absurd. It does neatly get rid of the question of what things really are. But does it make sense for two things to have a relation — to interact — if they are nothing intrinsically?

It may be that all there is to existence is relationship. But if so, is there an insight yet to be had about how this can or must be the case?

These are questions that are too deep for me. Someone with a different training and temperament might be able to make progress on them, but not I. The one thing I can't do is dismiss the question of what the world really is by calling it an absurd question. Some advocates of science insist that questions science cannot answer are meaningless, but I find this unconvincing — and unattractively narrow-minded. The pursuit of science has led me to conclude that the future is open and novelty is real. Since I define science by adherence to an ethic rather than a method, I must accept the possibility of scientific methodologies that no one has yet conceived.

This brings us to the *really* hard problem: the problem of consciousness.

I get a lot of e-mails about consciousness. To most of them I reply that whereas there are real mysteries about consciousness, they're beyond what science can tackle with present knowledge. As a physicist, I have nothing to say about them.

There's only one person I allow to talk to me about the problem of consciousness — a dear friend named James George. Jim is a retired diplomat who was the Canadian high commissioner to India and Sri Lanka and ambassador to Nepal, Iran, and the Gulf States, among other countries. He is, I am told, legendary as an exponent of Canadian diplomacy in the age of prime ministers Pearson and Trudeau, when Canada spread the idea of peacekeeping to the world. Now in his nineties, he writes books on the spiritual foundations of environmental issues and helps run a foundation dedicated to environmental causes.[11] He is much admired by his wide circle of friends and acquaintances for his sage advice — and he is one of the few people I know who appears to live guided by a level of wisdom I cannot imagine attaining.

So when Jim says to me, "What you're telling me about the meaning of time in physics is fascinating, but you're leaving out the key element that all your thought points to, which is the role of consciousness in the universe," I listen. I listen, but I don't have much to say.

But at least I know roughly what he's talking about. Let me be clear about what I mean by the problem of consciousness. I don't mean the question of whether we could program a computer to know or reflect on its own state. Nor do I mean the question of how systems evolve from networks of chemical reactions to become autonomous agents, a term Stuart Kauffman uses to refer to systems that can make decisions to their own benefit. These are hard problems, but they appear to be solvable, scientific ones.

By *the problem of consciousness,* I mean that if I describe you in all the languages physical and biological sciences make available to us, I leave something out. Your brain is a vast and highly interconnected network of roughly 100 billion cells, each of which is itself a complex system running on controlled chains of chemical reactions. I could describe this in as much detail as I wanted, and I would never come close to explaining the fact that you have an inner experience, a stream of consciousness. If I didn't know, from my own case, that I'm conscious, my knowledge of your neural processes would give me no reason to suspect that you are.

What's most mysterious, of course, is not the content of our consciousness but the *fact* that we're conscious. Leibniz imagined shrinking himself down and walking around inside someone's brain the way he might walk around inside a mill (these days, we would say "a factory"). In the case of the mill, you could give a complete description of it by describing what a person walking around inside it would see. In the case of a brain, you couldn't.

One way to tell that something is left out of the physical description of the workings of the brain is to mention some questions that the physical description doesn't answer. You and I look at the woman in the red dress sitting at the next table. Do each of us experience the same sensation (of red, I mean)? Might it be that what you experience as red is the same sensation I experience as blue? How can we tell?

Suppose your sight was extended to the ultraviolet. What would the new colors look like? What would the raw sensation of them be?

What's missing when we describe a color as a wavelength of light or as certain neurons lighting up in the brain is the essence of the experience of perceiving red. Philosophers give these essences a name: *qualia.* The question is, Why do we experience the qualia of red when our eyes absorb photons of a certain wavelength? This is what the philosopher David Chalmers calls *the hard question of consciousness.*

Here's another way to ask it: Suppose we mapped the neuronal circuits in your brain onto silicon chips and uploaded your brain into a computer. Would that computer be conscious? Would it have qualia? Yet another question serves to focus our thoughts: Suppose you could do that without harming yourself. Would there now be two conscious beings with your memories whose futures diverge from there?

The problem of qualia, or consciousness, seems unanswerable by science because it's an aspect of the world that is not encompassed when we describe all the physical interactions among particles. It's in the domain of questions about what the world really is, not how it can be modeled or represented.

Some philosophers argue that qualia simply are identical to certain neuronal processes. This seems to me wrong. Qualia may very well be correlated with neuronal processes but they are not the same as neuronal processes. Neuronal processes are subject to description by physics and chemistry, but no amount of detailed description in those terms will answer the questions as to what qualia are like or explain why we perceive them.

I don't doubt that there's a lot we could learn about the relationship between qualia and the brain that would take us closer to formulating the problems of consciousness and qualia as scientific questions. We can do experiments on conscious subjects that might teach us a lot about exactly what features or aspects of neuronal processes are associated with qualia. These are scientific questions, amenable to the methodologies of science.

The questions about qualia make consciousness a genuine mystery, one that hasn't yet been penetrated by the methods of science. I don't know whether it ever will be. Perhaps when we know a lot more about biology and the brain, it will lead to a revolutionary transformation in the language we use to describe living and thinking animals. After that revolution, we may have concepts and language unimaginable to us now that will let us formulate the mysteries of consciousness and qualia as scientific questions.

The problem of consciousness is an aspect of the question of what the world really is. We don't know what a rock really is, or an atom, or an electron. We can only observe how they interact with other things and thereby describe their relational properties. Perhaps everything has external and internal aspects. The external properties are those that science can capture and describe — through interactions, in terms of relationships. The internal aspect is the intrinsic essence; it is the reality that is not expressible in the language of interactions and relations. Consciousness, whatever it is, is an aspect of the intrinsic essence of brains.

One further aspect of consciousness is the fact that it takes place in time. Indeed, when I assert that it is always some time in the world, I am extrapolating from the fact that my experiences of the world always take place in time. But what do I mean by my experiences? I can speak about them scientifically as instances of recordings of information. To speak so, I need not mention consciousness or qualia. But this may be an evasion, because these experiences have aspects that are consciousness of qualia. So my conviction that what is real is real in the present moment is related to my conviction that qualia are real.

One way to put this last point, and to frame the whole argument of this book, is in terms of the philosophy called naturalism. This is the view that all that exists is the natural world that science describes. After the arguments of this book it becomes apparent that there are two very different versions of a naturalistic philosophy which are distinguished by the role that time is thought to play in the natural world. Here I have advocated a view that might be called *temporal naturalism*, according to which all that exists, exists at a moment of time, which is one of a succession of moments. This is opposed to a view that might be called *timeless naturalism*, which holds that the experience of the present moment and the flow of time are illusions rather than primary aspects of the natural world. Instead, timeless naturalism claims the natural world really consists of a timeless mathematical object from which all else is emergent.

From the point of view developed in this book, timeless naturalism may be said to have "looking-glassed" the term naturalism by taking a construction of our imaginations that is not part of nature and substituting it for the natural world. (The philosopher Galen Strawson uses the term looking-glassed to describe a move in a philosophical argument where a position is claimed to substitute for its opposite.) According to temporal naturalism, nothing exists outside of time, while according to timeless naturalism, all that exists is outside of time. I would claim that temporal naturalism offers a better basis for the kind of naturalism Strawson and Thomas Nagel hope for in which qualia are an intrinsic part of the natural world.

◆

Science is one of the great human adventures. The growth of knowledge is the spine of any telling of the human story, and for those fortunate enough to participate in it, it is the core of our lives. While its future is unpredictable — otherwise there would be no research — the only certainty is that we will know more in the future. For on every scale, from an atom's quantum state to the cosmos, and at every level of complexity, from a photon made in the early universe and winging its way toward us to human personalities and societies, the key is time and the future is open.

NOTES

1. This book can be taken as an introduction to, or popularization of, a rigorously argued work in natural philosophy I am collaborating on with Roberto Mangabeira Unger — which will result in a book tentatively titled *The Singular Universe and the Reality of Time,* in which we advocate the reality of time and the evolution of laws and examine possible resolutions of what we have called the *meta-laws dilemma* (see chapter 19).

2. Earlier versions of the arguments presented here may be found in the following papers, as well as in research publications listed in the notes below:

   Lee Smolin, "A Perspective on the Landscape Problem," arXiv:1202.3373v1 [physics. hist-ph] (2012);

   ———, "The Unique Universe," *Phys. World,* June 2, 21-6 (2009);

   ———, "The Case for Background Independence," in *The Structural Foundations of Quantum Gravity,* ed. Dean Rickles *et al.* (New York: Oxford University Press, 2007);

   ———, "The Present Moment in Quantum Cosmology: Challenges for the Argument for the Elimination of Time," in *Time and the Instant,* ed. Robin Durie (Manchester, U.K.: Clinamen Press, 2000);

   ———, "Thinking in Time Versus Thinking Outside of Time," in *This Will Make You Smarter,* ed. John Brockman (New York: Harper Perennial, 2012);

   Stuart Kauffman & Lee Smolin, "A Possible Solution to the Problem of Time in Quantum Cosmology," arXiv:gr-qc/9703026v1 (1997).

1. This view diminishes more than time, for it reduces all aspects of our experience of the world — color, touch, music, emotions, complex thoughts — to rearrangements of atoms. This is the core of the atomists' view of the world proposed by Democritus and Lucretius, formalized in John Locke's "primary and secondary qualities," and seemingly confirmed by every aspect of the progress of science since. In this view, what is real is motion — in the modern conception, transitions among quantum states. All else is to some extent illusion. My purpose is not to challenge any

of this wisdom, much of which must be taken as true, so well supported is it by science. My aim is only to challenge the last step, which claims that time, too, is an illusion.

2. The only exception, as we will see developed in chapter 11, is if our universe can be argued to be a typical member of the collection of universes.

3. Some readers will immediately ask whether there must be laws that govern the evolution of laws. This leads to the meta-laws problem, discussed at length in chapter 19.

4. Charles Sanders Peirce, "The Architecture of Theories," *The Monist*, 1:2, 161–76 (1891).

5. Roberto Mangabeira Unger, *Social Theory: Its Situation and Its Task*, vol. 2 of *Politics*, (New York: Verso, 2004), pp. 179–80.

6. Paul A. M. Dirac, "The Relation Between Mathematics and Physics," *Proc. Roy. Soc.* (Edinburgh) 59: 122–29 (1939).

7. Quoted in James Gleick, *Genius: the Life and Science of Richard Feynman* (New York: Pantheon, 1992), p. 93.

8. "Richard Feynman — Take the World from another Point of View," *NOVA* (PBS, 1973). Transcript at http://calteches.library.caltech.edu/35/2/PointofView.htm.

9. The first publication of this idea was Lee Smolin, "Did the Universe Evolve?" *Class. Quantum. Grav.* 9: 173–91 (1992).

10. "Dynamical" is a word I use often in this book. It means changeable, subject to law.

1. FALLING

1. This is despite many serious attempts by Islamic and medieval philosophers to understand the causes of motion.

2. Mathematicians like to speak of curves, numbers, and so forth as mathematical "objects," which implies a kind of existence. If you aren't comfortable adopting a radical philosophical position by a habit of language, you might want to call them concepts instead. I will use either word interchangeably when discussing mathematics, so as not to prejudice the question of what kind of existence they have.

3. It's also not *quite* true to say that the truths of mathematics are outside time, since, as human beings, our perceptions and thoughts take place at specific moments in time — and among the things we think about in time are mathematical objects. It's just that those mathematical objects don't seem to have any existence in time themselves. They are not born, they do not change, they simply are.

4. Many other great mathematicians believe this, such as Alain Connes. See Jean-Pierre Changeux & Alain Connes, *Conversations on Mind, Matter, and Mathematics*, ed. & trans. M. B. DeBevoise (Princeton, NJ: Princeton University Press, 1998).

2. THE DISAPPEARANCE OF TIME

1. One wonders whether anyone in antiquity noticed that water from a fountain traces a parabolic path. There are Greek vases which show water from fountains falling along what look like parabolas, so it would not have been impossible for a mathematician to have observed that and wondered whether falling bodies generally trace parabolas.

2. Aristotle, *On the Heavens*, Book 1, chapter 3.
3. I know several mathematicians and physicists who had to choose between a career in music and science. One is João Magueijo, who was trained as a composer of contemporary classical music before he decided to switch to physics. Being a person of extremes, he says he has not touched a piano since. Knowing him helps me to imagine the character of Galileo.
4. More precisely, the epicycle rotates with an Earth year for the outer planets: Mars, Jupiter, and Saturn. A different arrangement is necessary for Mercury and Venus.
5. Image from: Peter Apian, *Cosmographia* (1539). Reprinted in Alexandre Koyre, *From the Closed World to the Infinite Universe* (Baltimore, MD; Johns Hopkins, 1957).
6. This is of course a considerable simplification of a complex story of a turbulent relationship between mentor and assistant.
7. As suggested in *Agora*, a 2009 film by the Spanish-Chilean director Alejandro Amenábar.
8. When Newton presented the consequences of his laws of motion in his *Principia Mathematica*, he used more basic mathematics, rather than the calculus he had invented long before. This appears mysterious until you realize that he hadn't yet published the calculus; thus he had to explain his discoveries in mathematics his readers would have known.
9. Consider a ball falling near the surface of the Earth. It's pulled by the gravity of every atom making up the Earth. A key insight of Newton's was that all these forces can be added together, and the result is as if there were a single object pulling on the ball from the center of the Earth. If I throw the ball up, that distance may increase by a few meters, which is a very small change indeed, so the force hardly changes at all. The force on an object thrown or dropped can be taken to be constant. This implies that the acceleration is constant, which was Galileo's great discovery.

3. A GAME OF CATCH

1. Some would object that mathematics can code time — i.e., f(t) is a function of time. This completely misses the point, which is that the function f(t) is timeless.

4. DOING PHYSICS IN A BOX

1. Sara Diamond et al., *CodeZebra Habituation Cage Performances* (Rotterdam: Dutch Electronic Arts Festival, 2003).
2. Thanks to Saint Clair Cemin for discussions on this.
3. Consider a system of stars moving under their mutual gravitational influence. The interaction of two stars can be described exactly; Newton solved that problem. But there is no exact solution to the problem of describing the gravitational interaction of three stars. Any system of three or more bodies must be treated approximately. Such systems show a wide range of behaviors, including chaos and an extreme sensitivity to initial conditions. Although this is the next simplest system to the two-star problem, which Newton solved in the 17th century, these phenomena were not discovered until early in the 20th, by the French mathematician Henri Poincaré. Comprehending the so-called three-body problem required the invention of a whole new branch of mathematics: chaos theory. More recently, systems of thousands or mil-

lions of bodies can be treated in simulations carried out on supercomputers. These have given us insights into the behavior of stars in galaxies and even the interactions of galaxies in clusters. But the results gained, while useful, are based on the roughest of approximations. Stars consisting of vast numbers of atoms are treated as if they were points, and the influence of anything outside the system is usually ignored.

5. THE EXPULSION OF NOVELTY AND SURPRISE

1. Where we will raise and explain the apparent paradox that the laws of thermodynamics, such as the law that entropy can only increase, are not reversible in time, whereas the more fundamental laws are reversible.
2. Ludwig Boltzmann, *Lectures on Gas Theory* (Dover Publications, 2011).

6. RELATIVITY AND TIMELESSNESS

1. *The Principle of Relativity* (Dover Publications, 1952), consisting of seven of Einstein's papers, two by Hendrik Antoon Lorentz, and one each by Hermann Weyl and Hermann Minkowski.
2. "On the Electrodynamics of Moving Bodies," *Ann. der Phys.* 17(10): 891–921; "Does the Inertia of a Body Depend upon Its Energy Content?" *Ann. der Phys.* 18: 639–41 (1905).
3. Those readers who want to see the arguments thus explained can consult the on-line appendices at www.timereborn.com.
4. Strictly speaking, it is not necessary to assume that the speed of light is a speed limit, but it makes the pedagogy much simpler.
5. This isn't the same as saying that there is a fact about whether two events are simultaneous but it's impossible to know what it is. Because different observers will disagree about whether two events are simultaneous, there is no objective meaning to saying they are or aren't.
6. This does *not* mean that all clocks will tick the same number of times between two events. Consider two moving clocks that pass each other when they both read noon, then separate. One of them accelerates and reverses direction, passing the other clock again when that clock reads 12:01. The accelerating clock will display a different time. But the point is that all observers will agree about how many times one particular clock ticks between events. The clock that ticks the most times between two events is the one that is free-falling — and because the time a free-falling clock measures is distinguished in this way, we call it the proper time.
7. Hermann Weyl, *Philosophy of Mathematics and Natural Science* (Princeton, NJ: Princeton University Press, 1949).
8. If the region of spacetime is limited in space, you can also get from any A to any B in A's causal future by a series of steps using several intermediate X's. So the infinite extent of Minkowski spacetime helps make the argument in a single elegant step, but it's not essential.
9. Hilary Putnam, "Time and Physical Geometry," *Jour. Phil.* 64: 240–47 (1967).
10. John Randolph Lucas, *The Future* (Oxford, U.K.: Blackwell, 1990) p. 8.
11. The geodesics of spacetime, as opposed to space, are the paths that take the most

proper time rather then the shortest distance. This is a quirk of the way spacetime geometry is formulated; a free-falling clock ticks faster and thus more often than any other clock traveling between two events. This leads to a good piece of advice: If you want to stay young, accelerate.

12. The technical name for this property is *general coordinate invariance*; it is closely related to another property called *diffeomorphism invariance.* Newtonian mechanics can also be formulated in a way in which the clock can be part of the system and there is complete freedom to specify it. This formulation was developed by Julian Barbour in collaboration with Bruno Bertotti. It goes some way toward making the Newtonian paradigm relational, but it still is based on timeless laws acting on a timeless configuration space.

7. QUANTUM COSMOLOGY AND THE END OF TIME

1. Charles W. Misner, Kip S. Thorne, & John Archibald Wheeler, *Gravitation* (San Francisco: W. H. Freeman, 1973).

2. More on the different interpretations of quantum theory and their implications for the arguments of this book can be found in the on-line appendices.

3. The quantum state yields these probabilities via a two-step process. In the first step, the quantum state can be represented by giving a number for every possible configuration, called the *quantum amplitude* for that configuration. In the second step, you take the square of the amplitude of each configuration to get the probability that the system is in that configuration. Why these two steps? The amplitude is a complex number—a combination of two ordinary real numbers. This coding allows probability distributions for other quantities, such as momentum, to be coded into the same quantum state.

4. So if you want to check the prediction coming from a quantum state for the probabilities of finding the atom's electrons in various places, you prepare many atoms in that state and measure the positions of the electrons in each atom. Summing these gives you an experimental probability distribution. You can compare the experimental probability with the theoretical one computed from the quantum state. If they agree to within a reasonable margin of error, you have evidence that the initial assertion that the system was in a particular quantum state is correct.

5. The constant of proportionality is h, the famous Planck constant, denoting the value of a quantum of energy and named for Max Planck, its discoverer.

6. There are approximate descriptions of quantum cosmological states corresponding to expanding universes but these rely on extremely delicate choices of initial conditions. The generic state is a superposition of expanding and contracting universes. I should also mention that this isn't the only argument for the elimination of time in quantum cosmology, but it suffices for our purposes. Other arguments are given in the context of path-integral approaches to quantum gravity; also, Connes and Rovelli propose that time emerges as a consequence of the universe's having a finite temperature.

7. Yet another problem arises from the fact that in quantum mechanics, not all properties that can be observed have definite values at all times. So not all quantum states of a system have definite values of the system's energy, but some of them do. These states of definite energy, it turns out, also vibrate with a definite frequency. Indeed,

that's all they're doing — vibrating in place with a frequency proportional to the system's energy.

For many systems, there is a discrete set of states with definite energy. We say that the energy of these systems is quantized. But most quantum states do not have definite values of energy; in such a state, there are probabilities for the system to have different energies. Systems in these states also lack definite values of frequency.

To get a quantum system to do more than simply oscillate in place, you have to put it into a state without a definite energy value. This is easy to do, because of a principle known as the *superposition principle*, which says that quantum states can be added up. This is an aspect of the wave properties of a quantum system: A guitar or piano string vibrates at several frequencies simultaneously, and the string's motion is the sum of oscillations at each individual frequency. Throw two stones into a bucket of water: Each generates a wave, and the pattern on the water when they meet is the sum of the patterns made by each individual splash. The superposition principle works like that; given any two quantum states, you can make a third by adding them up.

This ability to add up quantum states is essential to our argument that Newtonian physics approximates quantum mechanics. We need it to reproduce the simple fact that in Newtonian physics configurations change as particles move around in space. This cannot be deduced from states that just oscillate in time, as do states of definite energy. To reproduce motion, we must have states whose behavior is more complex, and this requires states with indefinite energy values. These are built by adding up, or superposing, states with different energies.

But in quantum cosmology, all states have the same energy, so the usual way of extracting ordinary motion from quantum physics fails. We cannot deduce the predictions of general relativity from the quantum state of the universe.

8. Abhay Ashtekar, "New Variables for Classical and Quantum Gravity," *Phys. Rev. Lett.* 57:18, 2244–47 (1986).

9. Ted Jacobson & Lee Smolin, "Nonperturbative Quantum Geometries," *Nucl. Phys. B.*, 299:2, 295–345 (1988).

10. Carlo Rovelli & Lee Smolin, "Knot Theory and Quantum Gravity," *Phys. Rev. Lett.* 61:10, 1155–58 (1988).

11. Thomas Thiemann, "Quantum Spin Dynamics (QSD): II. The Kernel of the Wheeler–DeWitt Constraint Operator," *Class. Quantum Grav.* 15, 875–905 (1998).

12. Recently developed quantum-cosmology models study the quantum versions of simplified cosmological models like those we discussed in chapter 6. These are called loop-quantum-cosmology models. Earlier quantum-cosmology models were studied with crude approximations that muddied the fundamental issues; the recent models are simple and precisely defined enough to have yielded exact solutions to these equations. Impressive as this is, it must be emphasized that they are vastly simplified models. In particular, the problem of time is sidestepped, by speaking not of time but of correlations between the values of different observables. One field is treated as a clock with respect to which the changes in the other fields are measured. This provides an approximate, and relational, approach to extracting time from a timeless description of the world. Moreover, the issues are not limited to loop quantum gravity or loop quantum cosmology, even if they are most urgent in those contexts. String theory, to the extent that it can be applied to a closed cosmological

context, has an analog of a Wheeler-DeWitt equation. And some of the speculation about infinite universes, eternal inflation, and the like is set in the context of the Wheeler-DeWitt equations. The problems of interpreting the timeless universe that results is a challenge for all theorists who think about unification or the very early universe.

INTERLUDE: EINSTEIN'S DISCONTENT

1. Jim Brown tells me that Carnap had in mind something like the distinction between primary and secondary quantities. We experience red, but what is really happening is that atoms are vibrating and giving off light at a certain frequency. We experience time passing, but what is really true is that we are a bundle of world lines in a block universe, with the ability to perceive and store memories. To me, this is a way of stating the issue but does not resolve it.
2. *The Philosophy of Rudolf Carnap: Intellectual Autobiography*, ed. Paul Arthur Schillp (La Salle, IL: Open Court, 1963) pp. 37–8.

8. THE COSMOLOGICAL FALLACY

1. Carlo Rovelli, *The First Scientist: Anaximander and His Legacy* (Yardley, PA: Westholme Publishing, 2011).
2. Andrew Strominger, "Superstrings with Torsion," *Nucl. Phys. B* 274:2, 253–84 (1986).
3. A dilemma is an argument leading to a choice of two conclusions, neither of which is acceptable.
4. Someone might object that when we construct cosmological models in general relativity, we do apply Einstein's equations to the whole universe. But we don't. What we apply is a truncation of Einstein's equations to a subsystem consisting of the radius of curvature of the universe. Anything small — including us, the observers — is cut out of the system modeled.
5. For example, the Standard Model could be enhanced by adding extremely massive particles, which would hardly affect the universe for most of its history.

9. THE COSMOLOGICAL CHALLENGE

1. Other fixed-background structures include the geometry of the spaces where quantum states live; the notion of distance on those spaces, used to define probabilities; and the geometry of the spaces where the degrees of freedom of the Standard Model live. Background structures used in general relativity include the differential structure of spacetime and, often, the geometry of asymptotic boundaries.
2. The terms *background-dependent* and *background-independent* have a narrower use in discussions of quantum theories of gravity; in that context, a background-dependent theory is one that presumes a fixed background of classical spacetime. Perturbative theories, such as perturbative quantum general relativity and perturbative string theory, are background-dependent. Background-independent approaches to quantum gravity include loop quantum gravity, causal sets, causal dynamical triangulations, and quantum graphity.
3. Amit P. S. Yadav & Benjamin Wandelt, "Detection of Primordial Non-Gaussianity

(fNL) in the WMAP 3-Year Data at Above 99.5% Confidence," arXiv:0712.1148 [as-tro-ph], PRL100,181301, 2008.

4. Xingang Chen *et al.*, "Observational Signatures and Non-Gaussianities of General Single Field Inflation," arXiv:hep-th/0605045v4 (2008); Clifford Cheung *et al.*, "The Effective Field Theory of Inflation," arXiv.org/abs/0709.0293v2 [hep-th] (2008); R. Holman & Andrew J. Tolley, "Enhanced Non-Gaussianity from Excited Initial States," arXiv:0710.1302v2 (2008).

5. This does not mean that effects of initial conditions on the CMB can never be distinguishable from changes in the inflationary theory, at least within fixed classes of models. See Ivan Agullo, Jose Navarro-Salas, Leonard Parker, arXiv:1112.1581v2. Many thanks to Matthew Johnson for discussions on this point.

6. The uniqueness of the universe bedevils other attempts to test theories of the early universe. In ordinary laboratory physics, we have always to deal with noise arising from statistical uncertainties in the data. This can often be reduced by making many measurements, because the effect of random noise decreases as more trials are averaged together. Because the universe happens only once, this is impossible in some cosmological observations. These statistical uncertainties are known as *cosmic variance.*

7. Lee Smolin, "The Thermodynamics of Gravitational Radiation," *Gen. Rel. & Grav.* 16:3, 205–10 (1984); "On the Intrinsic Entropy of the Gravitational Field," *Gen. Rel. & Grav.* 17:5, 417–37 (1985).

8. Maybe a phase transition will get us, as the false vacuum that we may be living in decays. See, for example, Sidney Coleman & Frank de Luccia, "Gravitational Effects on and of Vacuum Decay," *Phys. Rev. D* 21:12, 3305–15 (1980).

9. This, by the way, explains why falling bodies travel along parabolas — those curves satisfy equations which are simple because they require only two pieces of data to define them, which are the acceleration due to gravity and the initial speed and direction of motion.

10. PRINCIPLES FOR A NEW COSMOLOGY

1. Here I'm following the advice of David Finkelstein, emeritus professor at Georgia Tech and one of the sages of contemporary physics, who once told me that a running start for the big conceptual leap we need in physics is afforded by contemplating its history over the past four centuries.

2. Be careful to distinguish a symmetry from a gauge symmetry. The first entails physical transformations that leave the laws unchanged. The second is a mathematical rewriting of the description of a system's configuration. The argument I give here rules out the first but not the second.

3. E. Noether, "Invariante Variationsprobleme," *Nachr. v. d. Ges. d. Wiss. zu Göttingen,* pp. 235–57 (1918).

4. This general reasoning is confirmed within general relativity theory, which, when applied to a closed universe, has neither symmetries nor conservation laws.

5. Roger Penrose was arguing this a long time ago. Indeed, we see from the example of string theory that the more symmetry a theory has, the less its explanatory power.

6. The only thing in Peirce's conclusion that is not precise is what he meant by evolution. Scholars have argued that he was referring to something like Darwinian natu-

ral selection. It is known that he was very much influenced by Darwin. But from the text alone, we can assume only that he meant evolution in the more general sense of changing in time according to some dynamical process. This is enough for our present argument, which is to establish that the *Why these laws?* question can be explained scientifically only if time is real.

7. Roberto Mangabeira Unger, draft manuscript.

11. THE EVOLUTION OF LAWS

1. Lee Smolin, "Did the Universe Evolve?" *Class. Quant. Grav.* 9: 173–91 (1992).

2. Alex Vilenkin, "Birth of Inflationary Universes," *Phys. Rev. D,* 27:12, 2848–55 (1983); Andrei Linde, "Eternally Existing Self-Reproducing Chaotic Inflationary Universe," *Phys. Lett. B,* 175:4, 395–400 (1986).

3. Several critiques of cosmological natural selection were published, and to my knowledge all were answered in the appendix to *The Life of the Cosmos* and subsequent papers. For the critiques, see T. Rothman and G.F.R. Ellis, "Smolin's Natural Selection Hypothesis," *Q. Jour. Roy. Astr. Soc.* 34, 201–12 (1993); Alex Vilenkin, "On Cosmic Natural Selection," arXiv:hep-th/0610051v2 (2006); Edward R. Harrison, "The Natural Selection of Universes Containing Intelligent Life," *Q. Jour. Roy. Astr. Soc.* 36, 193–203 (1995); Joseph Silk, "Holistic Cosmology," *Science,* 277:5326, 644 (1997); and John D. Barrow, "Varying G and Other Constants," arXiv:gr-qc/9711084v1 (1997). In particular, the claim that there's an easy argument that changing Newton's constant (fixing all other parameters) increases the numbers of black holes is wrong, because the complicated effects on galaxy and star formation and stellar evolution are not taken into account.

4. In biological evolution, there are actually two landscapes: the landscape of genes, which describes possible genotypes (DNA sequences), and the landscape of phenotypes, which are the physical expression of the genes. In applying natural selection to physics, you also have two levels of description. The probability of a universe reproducing depends on the values of the parameters of the Standard Model — these are analogous to the phenotypes. But in a fundamental theory like string theory, the Standard Model is an approximate description; underlying it is a choice of theories — these are analogous to the genotypes. In biological evolution, the relation between genotype and phenotype can be complex and indirect, and the same is true in physics. Thus, to be careful, you must distinguish between the landscape of a proposal for a fundamental theory, such as string theory, and the landscape of parameters of the Standard Model.

5. Others are (1) a reversal of the sign of the proton/neutron mass difference; (2) an increase or decrease in the Fermi constant large enough to affect the energy and matter ejected by supernovas; (3) an increase in the neutron/proton mass difference, the electron mass, the electron/neutrino mass, and the fine structure constant, or a decrease in the strong-interaction coupling large enough to destabilize carbon (or any simultaneous change having the same effect); and (4) an increase in the mass of the strange quark.

6. James M. Lattimer & M. Prakash, "What a Two Solar Mass Neutron Star Really Means," arXiv:1012.3208v1 [astro-ph.SR] (2010).

7. In the original paper on cosmological natural selection, as well as in *The Life of the*

Cosmos, I used the lower estimate for the critical mass — that is, 1.6 solar masses. When I learned about the observation of a neutron star with twice the solar mass, I started to write a paper pointing out that cosmological natural selection had been falsified. I was looking forward to this, because the second best thing that can happen in the field of quantum gravity is to make a prediction that is refuted by an experiment. However, I looked again into the theoretical estimates for the critical mass and found that the experts cautioned that these would still allow a 2-solar-mass kaon-neutron star.

8. See A. D. Linde, *Particle Physics and Inflationary Cosmology* (Chur, Switzerland: Harwood, 1990), pp. 162–8, especially the argument leading to eq. 8.3.17. (The book is also available at arXiv:hep-th/0503203v1.) The parameter that can raise the density fluctuations is the strength by which the inflaton (the particle bearing the inflationary force) interacts. As Linde shows, in some simple models, raising this parameter decreases the size of the universe by the exponential of the inverse square root of that interaction parameter. Many thanks to Paul Steinhardt for a conversation clarifying this issue.

9. For more details on cosmological natural selection, see *The Life of the Cosmos* or my following papers: "The Fate of Black Hole Singularities and the Parameters of the Standard Models of Particle Physics and Cosmology," arXiv:gr-qc/9404011v1 (1994); "Using Neutrons Stars and Primordial Black Holes to Test Theories of Quantum Gravity," arXiv:astro-ph/9712189v2 (1998); "Cosmological Natural Selection as the Explanation for the Complexity of the Universe," *Physica A: Statistical Mechanics and its Applications* 340:4, 705–13 (2004); "Scientific Alternatives to the Anthropic Principle," arXiv:hep-th/0407213v3 (2004); "The Status of Cosmological Natural Selection," arXiv:hep-th/0612185v1 (2006); and "A Perspective on the Landscape Problem", invited contribution for a special issue of *Foundations of Physics* titled "Forty Years Of String Theory: Reflecting On the Foundations," DOI: 10.1007/s10701-012-9652-x arXiv:1202.3373.

10. Roger Penrose has objected to me that black-hole singularities have a geometry very different from the initial cosmological singularity, making it unlikely that a black hole could be the source of our universe or any others. This is a concern, but the issue could be addressed if quantum effects played a big role in the elimination of the singularity.

11. Note that the idea of evolving laws does not, by itself, require global simultaneity. A change in laws could happen at an event that influences events only in its causal future. As explained in chapter 6, causal ordering is consistent with the relativity of simultaneity. But cosmological natural selection requires a global time to make sense — and this does indeed conflict with the relativity of simultaneity.

12. The justification for this is that the scale of the physics producing the bubbles is usually taken to be the grand unified scale, which is at least 15 orders of magnitude larger than the masses of the quarks and leptons of the Standard Model. Thus it's likely that these light fermion masses end up being essentially randomly chosen as bubble universes form.

13. B. J. Carr & M. J. Rees, "The Anthropic Principle and the Structure of the Physical World," *Nature* 278: 605–12 (1979); John D. Barrow & Frank J. Tipler, *The Anthropic Cosmological Principle* (New York: Oxford University Press, 1986).

14. Shamit Kachru *et al.,* "De Sitter Vacua in String Theory," arXiv:hep-th/0301240 v2 (2003).

15. Oliver DeWolfe *et al.,* "Type IIA Moduli Stabilization," arXiv:hep-th/0505160v3 (2005); Jessie Shelton, Washington Taylor, & Brian Wecht, "Generalized Flux Vacua," arXiv:hep-th/0607015 (2006).

16. George F. R. Ellis & Lee Smolin, "The Weak Anthropic Principle and the Landscape of String Theory," arXiv:0901.2414v1 [hep-th] (2009).

17. The universes with negative cosmological constant described by Washington Taylor and colleagues differ from ours in two respects. First, as is true in all string theories, there are extra dimensions involved. They are not observable, because they are tiny and curled up, but in Taylor's universes they can become large. This contradicts observations even more blatantly than having the wrong sign of the cosmological constant and might be taken as another wrong prediction of string theory. However, you can also argue that life could not exist in these worlds. Why that is, is not completely clear to me, because there are string-theory scenarios in which the particles and forces live on three-dimensional surfaces called branes, which float in extra dimensions. In that sort of configuration, life might be compatible with the extra dimensions being large.

    The hypothetical worlds with negative cosmological constant also have a symmetry our world does not, which is supersymmetry. This can prevent the formation of complex structure; however, it is possible that some fraction of these might allow supersymmetry to be spontaneously broken, in which case life might flourish there. As long as there are infinitely more string theories with negative cosmological constant than with positive, even if a very small fraction of the former can support life, these will dominate the latter. Thanks to Ben Freivogel for conversations on this issue.

18. At best, we could detect the influence of past collisions of other universes with our universe. This possibility has been studied, and it results in one-sided predictions — that is, something interesting might be seen that could be interpreted as the collision of other universes with our own, but if nothing is seen, as seems so far to be the case, no hypothesis is falsified. Stephen M. Feeney *et al.,* "First Observational Tests of Eternal Inflation: Analysis Methods and WMAP 7-Year Results," arXiv:1012.3667v2 [astro-ph.CO] (2011); and Anthony Aguirre & Matthew C. Johnson, "A Status Report on the Observability of Cosmic Bubble Collisions," arXiv:0908.4105v2 [hep-th] (2009) and 2011 *Rept. Prog. Phys.* 74:074901.

19. Steven Weinberg, "Anthropic Bound on the Cosmological Constant," *Phys. Rev. Lett.* 59:22, 2607–10 (1987).

20. In units of the Planck scale.

21. Adam G. Riess *et al.,* "Observational Evidence from Supernovae for an Accelerating Universe and a Cosmological Constant," *Astron. Jour.* 116, 1009–38 (1998).

22. One should beware, in evaluating the claim that Weinberg's argument provides evidence for the hypothesis that there exist other universes, of reasoning fallaciously that the fact that the cosmological constant takes on a very improbably small value is itself evidence for the assertion that ours is one of a vast collection of universes, in each of which the value of the cosmological constant is picked randomly. This reasoning is similar to that of the inverse gamblers fallacy discussed by the philosopher Ian Hacking. Suppose one walked into a room and saw someone rolling dice that

came up double sixes. One would be tempted to conclude that the dice had been rolled many times before, or were being rolled simultaneously many places, but these would be fallacious conclusions, because the probability of getting a double six is the same each time. Hacking calls this the inverse gamblers fallacy; Ian Hacking, "The Inverse Gambler's Fallacy: The Argument from Design. The Anthropic Principle Applied to Wheeler Universes." *Mind* 96:383 (July 1987), pp. 331–340. doi:10.1093/mind/XCVI.383.331. John Leslie objected in *Mind* 97:386 (April 1988), pp. 269–272. doi:10.1093/mind/XCVII.386.269 that the fallacy does not apply to the anthropic argument, because we must be in a universe hospitable to life. But Weinberg's argument correctly is not about hospitality but only about whether the universe is full of galaxies. We could live in a universe where only one galaxy had formed and still be alive — so the fact that the universe is full of galaxies is not necessary for life.

23. Jaume Garigga and Alex Vilenkin have pointed out, in "Anthropic Prediction for Lambda and the Q Catastrophe," arXiv:hep-th/0508005v1 (2005), that a particular combination of the two constants does better when applied to Weinberg's argument: It happens to be the cosmological constant divided by the fluctuation size cubed. But this leaves two issues: First, what sets the size of the fluctuations? Second, we already knew that the argument did all right when only the cosmological constant was considered. There are many combinations of the two constants that could be tried; the fact that one combination does better than the others is not surprising and, even if there is an argument for it, this does not constitute evidence for the hypothesis that our universe is one world of a vast multiverse.

24. Michael L. Graesser, Stephen D. H. Hsu, Alejandro Jenkins, & Mark B. Wise, "Anthropic Distribution for Cosmological Constant and Primordial Density Perturbations," hep-th/0407174, *Phys.Lett.* B600, 15–21 (2004).

25. An explanation for the value of the cosmological constant very different from Weinberg's is given by Rafael Sorkin and collaborators on the basis of the theory of causal sets: Maqbool Ahmed *et al.*, "Everpresent Lambda," arXiv:astro-ph/0209274v1 (2002).

12. QUANTUM MECHANICS AND THE LIBERATION OF THE ATOM

1. There are alternative views on quantum theory according to which it can be applied to the universe. For the reasons why I believe these fail, see the on-line appendices.

2. Momentum for ordinary particles is their mass times their velocity. Another expression of incompatible measurements is the uncertainty principle, which says that the more precisely position is measured, the less precisely we can measure momentum, and vice versa.

3. For a more technical explanation, see Lee Smolin, "Precedence and Freedom in Quantum Physics," arXiv:1205.3707v1 [quant-ph] (2012).

4. Charles Sanders Peirce, "A Guess at the Riddle," in *The Essential Pierce, Selected Philosophical Writings*, ed. Nathan Houser & Christian Kloesel (Bloomington IN: Indiana University Press, 1992), p. 277. Peirce's writings are rarely clear, so here's a summary from the *Stanford Encyclopedia of Philosophy* (http://plato.stanford.edu/entries/peirce/#anti):

One possible path along which nature evolves and acquires its habits was explored by Peirce using statistical analysis in situations of experimental trials in which the probabilities of outcomes in later trials are not independent of actual outcomes in earlier trials, situations of so-called "non-Bernoullian trials." Peirce showed that, if we posit a certain primal habit in nature, viz. the tendency however slight to take on habits however tiny, then the result in the long run is often a high degree of regularity and great macroscopic exactness. For this reason, Peirce suggested that in the remote past nature was considerably more spontaneous than it has now become, and that in general and as a whole *all* the habits that nature has come to exhibit have evolved. Just as ideas, geological formations, and biological species have evolved, natural habit has evolved.

5. John Conway & Simon Kochen, "The Free Will Theorem," *Found. Phys.*, 36:10, 1441 (2006).

6. For completeness, I should mention that some physicists respond to this argument by advocating a strong form of determinism, according to which the observers cannot be regarded as free to choose what to measure. From this "superdeterministic" point of view, we can imagine that there are correlations between choices observers make and choices atoms make, set far in the past of the experiment. Given this assumption, we can deny the conclusions of the theorems of Conway and Kochen, as well as Bell's theorem.

7. Lucien Hardy, "Quantum Theory from Five Reasonable Axioms," arXiv:quant-ph/0101012v4 (2001).

8. Lluis Masanes & Markus P. Mueller, "A Derivation of Quantum Theory from Physical Requirements," arXiv:1004.1483v4 [quant-ph] (2011). Related work was done by Borivoje Dakic & Caslav Brukner, "Quantum Theory and Beyond: Is Entanglement Special?" arXiv:0911.0695v1 [quant-ph] (2009).

9. Markus Müller has interesting work in progress relevant to this question.

13. THE BATTLE BETWEEN RELATIVITY AND THE QUANTUM

1. For a thorough discussion of de Broglie's work and an English translation of his 1927 paper, see Guido Bacciagaluppi & Antony Valentini, *Quantum Theory at the Crossroads: Reconsidering the 1927 Solvay Conference* (New York: Cambridge University Press, 2009), available at arXiv:quant-ph/0609184v2 (2009).

2. See John S. Bell, *Speakable and Unspeakable in Quantum Mechanics: Collected Papers on Quantum Philosophy* (New York: Cambridge University Press, 2004).

3. John von Neumann, *Mathematische Grundlagen der Quantenmechanik* (Berlin, Julius Springer Verlag, 1932), pp. 167 ff. or *Mathematical Foundations of Quantum Mechanics*, R. T. Beyer, trans. (Princeton, NJ: Princeton University Press, 1996).

4. Grete Hermann, "Die Naturphilosophischen Grundlagen der Quantenmechanik," *Abhandlungen der Fries'schen Schule* (1935).

5. David Bohm, *Quantum Theory* (New York: Prentice Hall, 1951).

6. ———, "A Suggested Interpretation of the Quantum Theory in Terms of 'Hidden' Variables. II," *Phys. Rev.*, 85:2, 180–93 (1952).

7. Antony Valentini, "Hidden Variables and the Large-scale Structures of Space-Time,"

in *Einstein, Relativity and Absolute Simultaneity*, eds. W. L. Craig & Q. Smith (London: Routledge, 2008), pp. 125–55.

8. Lee Smolin, "Could Quantum Mechanics Be an Approximation to Another Theory?" arXiv:quant-ph/0609109v1 (2006).

9. Albert Einstein, "Remarks to the Essays Appearing in This Collective Volume," in *Albert Einstein: Philosopher-Scientist*, ed. P. A. Schilpp (New York: Tudor, 1951), p. 671.

10. For a more technical explanation, see Lee Smolin, "A Real Ensemble Interpretation of Quantum Mechanics," arXiv:1104.2822v1 [quant-ph] (2011).

14. TIME REBORN FROM RELATIVITY

1. To be sure, the block-universe picture could incorporate the idea that laws change in time, but my claim is that it could not explain how and why they changed.

2. It might be thought that the aether was demolished by the Michelson-Morley experiment, but no one before Einstein in 1905 had the insight to recognize this.

3. The argument for this involves simple geometry, but I will not burden the reader with it. It can be found in any textbook on general relativity.

4. Suppose you're moving north with respect to this special observer. You will see the CMB radiation coming at you from the north to be blue-shifted as a consequence of the Doppler effect, which shifts the energy of each photon higher and increases the temperature of photons coming at you from the north. CMB photons coming from the south suffer the opposite effect; their frequencies are redshifted and their temperatures lower. So you can conclude that you're moving with respect to the cosmic microwave background. Conversely, the observer who sees the temperature to be the same in all directions can conclude that he is at rest with respect to the CMB.

5. In recent years, experiments have tested the validity of the principle of relativity in extreme circumstances, in which protons are observed to travel at .99999 of the speed of light. At this incredible speed, the effects of relativity are so important that the energy they carry is 10 billion times the energy inherent in their masses. I wouldn't have been surprised if those observations had revealed a breakdown in the principle of relativity, for such a breakdown is predicted by some approaches to quantum gravity at about those energies. Other recent observations test — and confirm — the principle that all photons have the same speed, to such an accuracy that the observations could have revealed whether one photon gains on another by a second after the pair has traveled together for 10 billion years. These results disappointed theorists who expected that quantum-gravity effects would alter the speed of light by a factor that depends on a photon's energy. Another set of observations confirmed to a high degree of accuracy that neutrinos have a speed limit the same as that of light (*pace* the premature reports of superluminal neutrinos that generated headlines around the world in 2011).

6. Other definitions for a preferred notion of time in general relativity have been proposed. Which one is correct is ultimately a scientific question to be decided by further developments, perhaps even by experiment. So we can suppose that there is a preferred notion of time, while leaving open the question of which one it is. Among the other proposals are: Chopin Soo & Hoi-Lai Yu, "General Relativity Without Paradigm of Space-Time Covariance: Sensible Quantum Gravity and Resolution of the Problem of Time," arXiv:1201.3164v2 [gr-qc] (2012); Niall Ó Murchadha, Cho-

pin Soo, & Hoi-Lai Yu, "Intrinsic Time Gravity and the Lichnerowicz-York Equation," arXiv:1208.2525vi [gr-qc] (2012); and George F. R. Ellis & Rituparno Goswami, "Space Time and the Passage of Time," arXiv:1208.2611v3 (2012).

7. Henrique Gomes, Sean Gryb, & Tim Koslowski, "Einstein Gravity as a 3D Conformally Invariant Theory," arXiv:1010.2481v2 [gr-qc] (2011).

8. This is known for technical reasons as the AdS/CFT correspondence.

9. For more about shape dynamics, see the on-line appendices.

10. Earlier in this chapter, I mentioned that some symmetric solutions to general relativity have a preferred state of rest and hence a preferred time. This is different. The earlier case is restricted to special solutions, whereas the preferred time identified by shape dynamics is general and exists even in spacetimes that have no symmetries. There is a weak restriction on the spacetime, which is that it have what is called a constant mean-curvature slicing; this is thought not to impede the application of the theory to cosmological spacetimes. This notion of time is global, and it is dynamically determined by the gravitational field and matter. So this is not a retreat to the absolute time of Newton. Roughly speaking, the chosen slices of spacetime are minimally curved. In the same sense that soap bubbles take shapes that minimize their curvature, the slices by which spacetime is divided up can minimize their curvature.

15. THE EMERGENCE OF SPACE

1. The architects, Saucier + Perrotte, did suggest, when we told them how much blackboard space we wanted, that the building could be covered entirely in slate and glass, so we could write all over it.

2. For a recent review, see J. Ambjørn et al., "Nonperturbative Quantum Gravity," arXiv:1203.3591v1 [hep-ph] (2012); "Emergence of a 4-D world from Causal Quantum Gravity", Phys. Rev. Lett. 93 (2004) 131301 [hep-th/0404156].

3. Fotini Markopoulou, "Space Does Not Exist, So Time Can," arXiv:0909.1861v1 [gr-qc] (2009).

4. Tomasz Konopka, Fotini Markopoulou, & Lee Smolin, "Quantum Graphity," arXiv:hep-th/0611197v1 (2006); Tomasz Konopka, Fotini Markopoulou, & Simone Severini, "Quantum Graphity: A Model of Emergent Locality," arXiv:0801.0861v2 (2008); Alioscia Hamma et al., "A Quantum Bose-Hubbard Model with Evolving Graph as Toy Model for Emergent Spacetime," arXiv:0911.5075v3 [gr-qc] (2010).

5. Petr Horava, "Quantum Gravity at a Lifshitz Point," arXiv:0901.3775v2 [hep-th] (2009).

6. T. Banks et al., "M Theory as a Matrix Model: A Conjecture," arXiv:hep-th/9610043v3 (1997).

7. Experts may point out that volume and area are not physical observables, because they are not invariant under spacetime diffeomorphisms. But there are cases where they are physical, either because they are properties of a boundary where diffeomorphisms are fixed or because a time gauge has been fixed, giving rise to a physical description of evolution generated by a hamiltonian.

8. See, for example, Aurelien Barrau et al., "Probing Loop Quantum Gravity with Evaporating Black Holes," arXiv:1109.4239v2 (2011).

9. In what time? Any definition of time! In loop quantum gravity time is arbitrary, as it

is a quantization of general relativity in which time can be chosen arbitrarily, reflecting its many-fingered nature.

10. In the original approach to loop quantum gravity, the graph is considered as contained in a three-dimensional space that has only the simplest properties. Nothing that can be measured—like length, area, or volume—is fixed. But the number of spatial dimensions is fixed, as is the space's connectivity, or topology. (By "topology," we mean the gross sense of how it hangs together; this is unchanged when the shape is seamlessly distorted.)

   Topology is best explained by examples and easiest to visualize in two dimensions. Consider a closed two-dimensional surface. It might be like a sphere or a torus (a doughnut shape). You can smoothly deform a sphere into a variety of shapes, but you cannot smoothly distort a sphere into a torus. Other topologies of two-dimensional surfaces can resemble doughnuts with many holes.

   Once we fix the topology of space, we can consider the various ways in which a graph can be embedded in it. For example, the graph's edges can be knotted up or braided or otherwise linked with one another. Each way of embedding the graph in space results in a distinct quantum state of geometry (although in most contemporary work in quantum gravity, the graphs are defined without reference to any embedding).

11. See, for example, Muxin Han & Mingyi Zhang, "Asymptotics of Spinfoam Amplitude on Simplicial Manifold: Lorentzian Theory," arXiv:1109.0499v2 (2011); Elena Magliaro & Claudio Perini, "Emergence of Gravity from Spinfoams," arXiv:1108.2258v1 (2011); Eugenio Bianchi & You Ding, "Lorentzian Spinfoam Propagator," arXiv:1109.6538v2 [gr-qc] (2011); John W. Barrett, Richard J. Dowdall, Winston J. Fairbairn, Frank Hellmann, Roberto Pereira, "Lorentzian Spin Foam Amplitudes: Graphical Calculus and Asymptotics," arXiv:0907.2440; Florian Conrady & Laurent Freidel, "On the Semiclassical Limit of 4d Spin Foam Models," arXiv:0809.2280v1 [gr-qc] (2008); Lee Smolin, "General Relativity as the Equation of State of Spin Foam," arXiv:1205.5529v1 [gr-qc] (2012).

12. Technically speaking, the dual of a triangulation of a 3-manifold.

13. See Fotini Markopoulou & Lee Smolin, "Disordered Locality in Loop Quantum Gravity States," arXiv:gr-qc/0702044v2 (2007).

14. This idea defines a research program I have worked on intermittently for years. See Markopoulou & Smolin, "Quantum Theory from Quantum Gravity," arXiv:gr-qc/0311059v2 (2004). See also: Julian Barbour & Lee Smolin, "Extremal Variety as the Foundation of a Cosmological Quantum Theory," arXiv:hep-th/9203041v1 (1992);

   Lee Smolin, "Matrix Models as Nonlocal Hidden Variables Theories," arXiv:hep-th/0201031v1 (2002);

   ———, "Quantum Fluctuations and Inertia," *Phys. Lett. A,* 113:8, 408–12 (1986);

   ———, "On the Nature of Quantum Fluctuations and Their Relation to Gravitation and the Principle of Inertia," *Class. Quant. Grav.* 3: 347–59 (1986);

   ———, "Stochastic Mechanics, Hidden Variables, and Gravity," in *Quantum Concepts in Space and Time,* ed. R. Penrose & C. J. Isham (New York: Oxford University Press, 1986);

   ———, "Derivation of Quantum Mechanics from a Deterministic Nonlocal Hidden

Variable Theory. 1. The Two-Dimensional Theory," IAS preprint, July 1983. http://inspirehep.net/record/191936.

15. Chanda Prescod-Weinstein & Lee Smolin, "Disordered Locality as an Explanation for the Dark Energy," arXiv:0903.5303v3 [hep-th] (2009).

16. Dark matter is a hypothetical kind of matter that gives off no light but is necessary if the rotations of galaxies are to be explained on the basis of Newton's laws.

17. Lee Smolin, "Fermions and Topology," arXiv:gr-qc/9404010v1 (1994).

18. C. W. Misner and J. A. Wheeler, *Ann. Phys. (U.S.A.)* 2, 525–603 (1957), reprinted in *Wheeler Geometrodynamics* (New York: Academic Press, 1962).

19. Fotini Markopoulou, "Conserved Quantities in Background Independent Theories," arXiv:gr-qc/0703027v1 (2007).

20. Francesco Caravelli & Fotini Markopoulou, "Disordered Locality and Lorentz Dispersion Relations: An Explicit Model of Quantum Foam," arXiv:1201.3206v3 (2012); Caravelli & Markopoulou, "Properties of Quantum Graphity at Low Temperature," arXiv:1008.1340v3 (2011); Caravelli *et al.*, "Trapped Surfaces and Emergent Curved Space in the Bose-Hubbard Model," arXiv:1108.2013v3 (2011); Florian Conrady, "Space as a Low-temperature Regime of Graphs," arXiv:1009.3195v3 [gr-qc] (2011). Another approach to geometrogenesis is in João Magueijo, Lee Smolin, & Carlo R. Contaldi, "Holography and the Scale-Invariance of Density Fluctuations," arXiv:astro-ph/0611695v3 (2006).

21. Graphs and triangulations are closely related. Given a triangulation, you can make a graph in which the nodes represent the tetrahedrons and two nodes are connected by an edge if the corresponding tetrahedrons are joined at a face.

22. The figure shows a quantum universe with one dimension of space and one of time, taken from R. Loll, J. Ambjørn, K. N. Anagnostopoulos, "Making the Gravitational Path Integral More Lorentzian, or: Life Beyond Liouville Gravity," arXiv:hep-th/9910232, *Nucl.Phys.Proc.Suppl.* 88, 241–244 (2000). Used with permission.

23. Alioscia Hamma *et al.*, "Lieb-Robinson Bounds and the Speed of Light from Topological Order," arXiv:0808.2495v2 (2008).

16. THE LIFE AND DEATH OF THE UNIVERSE

1. Richard Dawkins, *Climbing Mount Improbable* (New York: W. W. Norton, 1996).

2. A fluctuation is one of those words physicists use that can be confusing for lay readers. A fluctuation is a small random change in a small part of a system. A fluctuation can disorder a system, as when a drop of paint from a paintbrush ruins a carefully crafted portrait. But a fluctuation can also spontaneously lead to a higher degree of organization, as when a mutation resulting from a random change in a DNA molecule produces a fitter animal.

3. It's of interest to note that organic (or prebiotic) molecules have been detected not just on Earth but in meteorites, comets, and interstellar clouds of dust and gas.

4. Because the logarithm of 1 is zero. For technical reasons, we usually take the entropy to be the logarithm of the number of equivalent microstates.

5. "Über die von der molekularkinetischen Theorie der Wärme geforderte Bewegung von in ruhenden Flüssigkeiten suspendierten Teilchen," *Ann. der Phys.* 17 (8): 549–60 (1905).

6. Martin J. Klein, *Paul Ehrenfest: The Making of a Theoretical Physicist* (New York: Elsevier, 1970).

7. See, for example, *Time's Arrow*, by Martin Amis, or *The Curious Case of Benjamin Button*, a film based on the short story by F. Scott Fitzgerald.

8. Many thanks to Steven Weinstein of the University of Waterloo for discussions in which he convinced me of the importance of the electromagnetic arrow of time. His 2011 paper, "Electromagnetism and Time-Asymmetry," arXiv:1004.1346v2, strongly influenced the following section.

9. Roger Penrose, "Singularities and Time-Asymmetry," in S. W. Hawking & W. Israel, eds., *General Relativity: An Einstein Centenary Survey* (Cambridge, U.K.: Cambridge University Press, 1979), pp. 581–638.

10. Many physicists and philosophers have wondered whether there really are several distinct arrows of time. Might one or more arrows be explained by the others? The cosmological arrow of time is probably not related to the others.

   It's easy to imagine an expanding universe that expanded so fast that no gravitationally-bound structures would have time to form. Such a universe would remain in equilibrium forever and thus would not have a thermodynamic arrow of time. So the fact that the universe is expanding is not, by itself, sufficient to explain the thermodynamic arrow of time.

   It's also possible to imagine a universe that expands to its maximal size and then collapses. As far as we know now, this isn't the universe we live in, but there are solutions to the equations of general relativity that behave this way. This would be a world where the cosmological arrow of time reversed halfway through. Would the thermodynamic arrow reverse as well, so that all of a sudden spilt milk cleaned itself up and Humpty-Dumpty reassembled himself? Science-fiction writers like to imagine this, but it's wildly implausible.

   But the biological arrow of time may well be a consequence of the thermodynamic arrow. We age, it is claimed, because disorder accumulates in our cells. The thermodynamic arrow is also taken to explain at least some of the experiential arrow. We remember the past and not the future because memory is a form of organization, and organization decreases in the future — or so it is claimed.

   Finally, can the thermodynamic arrow of time be reduced to the choice of initial conditions? This was proposed by Penrose, who argued that his Weyl curvature hypothesis could explain the thermodynamic arrow of time because a universe with initially no black or white holes has much less entropy than it might if it were filled randomly with black and white holes. He relies here on the idea that black holes have entropy, an amazing fact uncovered by Jacob Bekenstein in 1972 and developed by Stephen Hawking soon afterward. Black holes have huge amounts of entropy, since the most irreversible thing you can do is to send something into a black hole. Taking into account the vast amount of entropy that might exist in all the black holes the universe might have started off with but didn't, the actual universe, without any initial black holes, started in a state of almost minimal entropy.

   Penrose's proposal succeeds as long as we retain the condition that the universe expands slowly and uniformly enough that gravitationally-bound structures can form. From this perspective, a complex universe is highly improbable, since most initial conditions would lead to a universe that started and stayed in equilibrium. It would be filled with light and gravity waves present from the beginning and carrying no images of the past or future. Black holes and white holes would dominate from the beginning. Within a world governed by time-symmetric laws, the explanation for why we live in a complex universe rests largely on the extremely improbable choice of time-asymmetric initial conditions.

11. The fundamental time-asymmetric law would have to lead to time-symmetric laws when approximated by an effective theory at low energy and away from regions of high spacetime curvature. Thus, the time asymmetry would be very pronounced in the very early universe, which would explain the need for highly time-asymmetric initial cosmological conditions.

12. Note that we are talking about properties of the whole universe, which are not properties of small subsystems of it. We can always apply probability to small subsystems or regions of the universe, but these do not exhaust all we want to know about the universe.

13. Of course, given infinite time, every scale of fluctuation happens infinitely many times. This makes it a bit tricky to say that rarer fluctuations happen fewer times, because the ratio of two infinite numbers is ill-defined.

17. TIME REBORN FROM HEAT AND LIGHT

1. The reader may ask whether Leibniz's principle of the identity of the indiscernibles is in contradiction with Bose statistics, which allows and encourages bosons to share the same quantum state. A brief response, expanded in the on-line appendices, is that Leibniz's principle forbids two events from having the same expectation values of quantum fields.

2. As I pointed out in chapter 10, this forbids the universe to be perfectly symmetric.

3. For more on self-organization see the books by Bak, Kauffman and Morowitz in the bibliography. One version of the principle of driven self-organization is the cycle theorem described in Morowtiz's book, another is the phenomena of self-organized criticality described in Bak's book.

4. Julian Barbour and Lee Smolin, "Variety, Complexity and Cosmology," hep-th /9203041.

5. Alan Turing, "The Chemical Basis of Morphogenesis," *Phil. Trans. Roy. Soc. Lond.* 237:641, 37–72 (1952).

18. INFINITE SPACE OR INFINITE TIME?

1. This is mind-boggling, but there's a simple argument for it. For details, see Brian Greene's latest book, *The Hidden Reality: Parallel Universes and the Deep Laws of the Cosmos* (New York: Knopf, 2011), or a discussion in the online appendices.

2. Imagine a flat two-dimensional plane. Pick a point, then pick a direction going outward from that point. That defines a line in the plane. Follow that line as far as it goes. It goes an infinite distance, but in the mind's eye of a mathematician it nonetheless goes somewhere. Where it goes is called *a point at infinity*. Pick another direction from the original point. You get another line. Follow that as far as it goes; it takes us to another point at infinity. The points at infinity make up a circle. The directions you can go from a point in a plane define a circle. Following those directions as far as they go, you reach the boundary of points at infinity. The same thing obtains

in flat three-dimensional space, except that the points at infinity make up a sphere. It also obtains if the space is infinite but negatively curved, like a saddle.

When you set out to solve the equations of general relativity, you have to specify information about what's happening at that boundary. You have to specify what is coming in from the boundary, and what is going out to it. The need to specify information about what's happening at the infinite boundary is not optional; it is required by the theory. (For the experts, the Einstein equations for a spatially infinite universe cannot be derived from a variational principle unless there are boundary terms added to the action and boundary conditions specified at spatial infinity.) You cannot describe what's in the universe without saying what's coming into and going out of the universe from the boundary. Even if the boundary is infinitely far away.

3. In the practice of general relativity, we often use spaces with infinite boundaries as convenient models of isolated systems. Consider a galaxy. In reality, it's a small part of the universe, but for some purposes we might want to model it as isolated; for example, we might want to model the interaction of the black hole in the center with stars in the galactic disk. So we draw a boundary around the galaxy and construct a solution of general relativity containing only what is within that boundary. But there are some technical hassles in dealing with information to be specified at a finite boundary. So, purely for technical convenience, we idealize the situation and push the boundary out to infinity. This greatly simplifies the description, because we can impose the condition that all the matter in that model is contained in the one galaxy. Nothing can come in or go out except gravitational waves and light, which we can use to observe the galaxy.

This kind of use of infinite spaces is pragmatic, and there can be no objection to it. The fact that information must be specified coming in from the infinite boundary reminds us that we're dealing with an idealization in which we cut out a part of the universe and describe it as if it were all there is. But it's nonsensical to model the whole universe as having an outer boundary, which requires the specification of information coming in from outside the infinite universe. Yet this is what we must do if we use general relativity as our cosmological theory and take the universe to be spatially infinite.

4. For more on these cyclic cosmologies, see Paul J. Steinhardt & Neil Turok, *Endless Universe: Beyond the Big Bang* (New York: Doubleday, 2007).

5. Martin Bojowald, "Isotropic Loop Quantum Cosmology," arXiv:gr-qc/0202077v1 (2002);

———, "Inflation from Quantum Geometry," arXiv:gr-qc/0206054vi (2002);

———, "The Semiclassical Limit of Loop Quantum Cosmology," arXiv:gr-qc /0105113v1 (2001);

———, "Dynamical Initial Conditions in Quantum Cosmology," arXiv:gr-qc /0104072v1 (2001); and

Shinji Tsujikawa, Parampreet Singh, & Roy Maartens, "Loop Quantum Gravity Effects on Inflation and the CMB," arXiv:astro-ph/0311015v3 (2004).

6. Jean-Luc Lehners, "Diversity in the Phoenix Universe," arXiv:1107.4551v1 [hep-ph] (2011).

7. Roger Penrose, *Cycles of Time: An Extraordinary New View of the Universe* (New York: Knopf, 2011).

8. Claims that circles have been detected are in:

V. G. Gurzadyan & R. Penrose, "CCC-Predicted Low Variance Circles in CMB Sky and LCDM," arXiv:1104.5675v1 [astro-ph.CO] (2011);

———, "More on the Low-Variance Circles in CMB Sky," arXiv:1012.1486v1 [astro -ph.CO] (2010);

———, "Concentric Circles in WMAP Data May Provide Evidence of Violent Pre– Big-Bang Activity," arXiv:1011.3706v1 [astro-ph.CO] (2010).

Several papers argue that this is consistent with noise:

I. K. Wehus & H. K. Eriksen, "A Search for Concentric Circles in the 7-Year WMAP Temperature Sky Maps," arXiv:1012.1268v1 [astro-ph.CO] (2010);

Adam Moss, Douglas Scott, & James P. Zibin, "No Evidence for Anomalously Low- variance Circles on the Sky," arXiv:1012.1305v3 [astro-ph.CO] (2011); and

Amir Hajian, "Are There Echoes from the Pre–Big Bang Universe? A Search for Low-Variance Circles in the CMB Sky," arXiv:1012.1656v1 (2010).

19. THE FUTURE OF TIME

1. This idea is realized in a model in Lee Smolin, "Matrix Universality of Gauge and Gravitational Dynamics," arXiv:0803.2926v2 [hep-th] (2008).
2. ———, "Unification of the State with the Dynamical Law, arXiv:1201.2632v1 [hep-th] (2012).
3. Wheeler also said, "No phenomenon is a real phenomenon until it is an observed phenomenon." I must say I come to appreciate his enigmatic and provocative challenges to us more and more as I mature.

EPILOGUE

1. For more on the view developed here and references, see Lee Smolin, "Time and Symmetry in Models of Economic Markets," arXiv:0902.4274v1 [q-fin.GN] (2009).
2. For an introduction to neoclassical economics, see Ross M. Starr, *General Equilib- rium Theory*, 2nd edition (New York: Cambridge University Press, 2011).
3. This is shown by the Sonnenschein-Mantel-Debreu theorem, or "anything goes the- orem," proved in 1972 by three highly influential economists. One of them is Hugo Sonnenschein, who is not just a member of the Chicago school of economists but served as president of that university. Hugo Sonnenschein, "Market Excess De- mand Functions," *Econometrica*, 40:3, 549–63 (1972). Debreu, G. "Excess Demand Functions," *Journal of Mathematical Economics* 1: 15–21 (1974), doi:10.1016/0304 -4068(74)90032-9; R. Mantel, "On the Characterization of Aggregate Excess De- mand," *Jour. of Econ. Theory* 7: 348–353 (1974), doi:10.1016/0022-0531(74)90100-8.
4. W. Brian Arthur, "Competing Technologies, Increasing Returns, and Lock-In by Historical Events," *Econ. Jour.* 99:394, 116–31 (1989).
5. Pia Malaney, "The Index Number Problem: A Differential Geometric Approach," Harvard PhD thesis, 1996.
6. Malaney and Weinstein's ideas prompted Samuel Vazquez, then a postdoc at Perim- eter Institute, to measure path dependence in real market data. What he was doing was impossible and heretical in the framework of neoclassical economics theory, but there it was, in real data, showing that the existence of funds with successful long-short arbitrage strategies proves that there is indeed curvature, and hence path

dependence, in the market. Samuel E. Vazquez & Simone Farinelli, "Gauge Invariance, Geometry and Arbitrage," arXiv:0908.3043v1 [q-fin.PR] (2009).

7. Vince Darley & Alexander V. Outkin, *A NASDAQ Market Simulation: Insights on a Major Market from the Science of Complex Adaptive Systems* (World Scientific, 2007).

8. I see the beginnings of this common conception in the fact that the theoretical biologist Stuart Kauffman and the philosopher of law Roberto Mangabeira Unger both speak of the need to formulate their domains in terms of the adjacent possible — the set of next steps — rather than in abstract timeless spaces of all possible configurations. .

9. The implications of these two principles are further developed in chapter 17 of my 2006 book, *The Trouble with Physics.*

10. Notice that relationships are exactly what mathematics expresses. Numbers have no intrinsic essence, nor do points in space; they are defined entirely by their place in a system of numbers or points — all of whose properties have to do with their relationships to other numbers or points. These relationships are entailed by the axioms that define a mathematical system. If there's more to matter than relationships and interactions, it is beyond mathematics.

11. James George is the author of *Asking for the Earth* (Barrytown NY: Station Hill Press, 2002) and *The Little Green Book on Awakening* (Barrytown NY: Station Hill Press, 2009). He is also a cofounder of the Threshold Foundation and president of the Sadat Peace Foundation, and he led the international mission to Kuwait to assess environmental damage in the wake of the Persian Gulf War.

BIBLIOGRAPHY

Here is a selection of mostly popular books on the topic of time in physics or cosmology (and related issues), many of them presenting alternative or conflicting ideas to those I have proposed in these pages.

Guido Bacciagaluppi & Antony Valentini, *Quantum Theory at the Crossroads: Reconsidering the 1927 Solvay Conference* (New York: Cambridge University Press, 2009).

Per Bak, *How Nature Works: The Science of Self-Organized Criticality* (New York: Copernicus, 1996).

Julian B. Barbour, *The End of Time: The Next Revolution in Physics* (New York: Oxford University Press, 2000).

——, *The Discovery of Dynamics: A Study from a Machian Point of View of the Discovery and the Structure of Dynamical Theories* (New York: Oxford University Press, 2001).

J. S. Bell, *Speakable and Unspeakable in Quantum Mechanics*, 2nd ed. (New York: Cambridge University Press, 2004).

James Robert Brown, *Platonism, Naturalism, and Mathematical Knowledge* (Oxford, U.K.: Routledge, 2011).

Bernard Carr, ed., *Universe or Multiverse?* (New York: Cambridge University Press, 2007).

Sean Carroll, *From Eternity to Here: The Quest for the Ultimate Arrow of Time* (New York: Dutton, 2010).

P.C.W. Davies, *The Physics of Time Asymmetry* (San Francisco: University of California Press, 1974).

David Deutsch, *The Fabric of Reality: The Science of Parallel Universes — and Its Implications* (New York: Allen Lane/Penguin Press, 1997).

Dan Falk, *In Search of Time: The History, Physics and Philosophy of Time* (New York: St. Martin's, 2010).

Adam Frank, *About Time: Cosmology and Culture at the Twilight of the Big Bang* (New York: Free Press, 2011).

Rodolfo Gambini & Jorge Pullin, *A First Course in Loop Quantum Gravity* (New York: Oxford University Press, 2011).

Marcelo Gleiser, *A Tear at the Edge of Creation: A Radical New Vision for Life in an Imperfect Universe* (New York: Free Press, 2010).

Brian Greene, *The Hidden Reality: Parallel Universes and the Deep Laws of the Cosmos* (New York: Knopf, 2011).

Stephen W. Hawking & Leonard Mlodinow, *The Grand Design* (New York: Bantam, 2010).

Stuart A. Kauffman, *At Home in the Universe: The Search for the Laws of Self-Organization and Complexity* (New York: Oxford University Press, 1995).

———, *The Origins of Order: Self-Organization and Selection in Evolution* (New York: Oxford University Press, 1993).

Helge Kragh, *Higher Speculations: Grand Theories and Failed Revolutions in Physics and Cosmology* (New York: Oxford University Press, 2011).

Janna Levin, *How the Universe Got Its Spots: Diary of a Finite Time in a Finite Space* (Princeton, NJ: Princeton University Press, 2002).

João Magueijo, *Faster than the Speed of Light: The Story of a Scientific Speculation* (Cambridge, MA: Perseus, 2003).

Roberto Mangabeira Unger, *The Self Awakened: Pragmatism Unbound* (Cambridge, MA: Harvard University Press, 2007).

Harold Morowitz, *Energy Flow in Biology*, (New York: Academic Press, 1968).

Richard Panek, *The 4-Percent Universe: Dark Matter, Dark Energy, and the Race to Discover the Rest of Reality* (Boston, MA: Houghton Mifflin Harcourt, 2011).

Roger Penrose, *Cycles of Time: An Extraordinary New View of the Universe* (New York: Knopf, 2011).

———, *The Road to Reality: A Complete Guide to the Laws of the Universe* (New York: Knopf, 2005).

———, *The Emperor's New Mind: Concerning Computers, Minds, and the Laws of Physics* (New York: Oxford University Press, 1989).

Huw Price, *Time's Arrow and Archimedes' Point: New Directions for the Physics of Time* (New York: Oxford University Press, 1996).

Lisa Randall, *Warped Passages: Unraveling the Mysteries of the Universe's Hidden Dimensions* (New York: Ecco/HarperCollins, 2005).

Carlo Rovelli, *The First Scientist: Anaximander and His Legacy* (Yardley, PA: Westholme Publishing, 2011).

Simon Saunders *et al.*, eds., *Many Worlds? Everett, Quantum Theory, and Reality* (New York: Oxford University Press, 2010).

Lee Smolin, *The Life of the Cosmos* (New York: Oxford University Press, 1997).

———, *Three Roads to Quantum Gravity* (New York: Basic Books, 2001).

———, *The Trouble with Physics* (Boston, MA: Houghton Mifflin Harcourt, 2006).

Paul J. Steinhardt & Neil Turok, *Endless Universe: Beyond the Big Bang* (New York: Doubleday, 2007).

Leonard Susskind, *The Cosmic Landscape: String Theory and the Illusion of Intelligent Design* (New York: Little, Brown, 2005).

Alex Vilenkin, *Many Worlds in One: The Search for Other Universes* (New York: Hill & Wang, 2006).

ACKNOWLEDGMENTS

The writing of this book has been a great adventure, reflecting a life-long engagement with the nature of time. As any traveler does, I owe an enormous debt to the many people who assisted, encouraged, guided, and sometimes led me on this journey.

The adventure began in 1980, when I was studying for the summer in Oxford as a guest of Roger Penrose. Roger told me that if I really wanted to think about the nature of time I had to talk with a fellow named Julian Barbour, who lived and worked in a village near Oxford. Arrangements were made, and along with the philosopher of science Amelia Rechel-Cohn, I paid him a visit. It was during those discussions that Julian began his philosophical mentorship of me, introducing me to Leibniz's writings and the ideas of relational space and time. I was one of the first, but far from the last, young physicists to have his head turned round and his thinking set in the right direction by Julian's tutelage.

The journey took an unexpected turn in 1986, when Andrew Strominger told me about his discovery of a great number of string theories and his worries that their abundance would defeat any attempt to derive the Standard Model of Particle Physics from pure principles. Meditating on this, I imagined a landscape of string theories similar to fitness landscapes in biology, in which a mechanism analogous to natural selection would govern the evolution of laws. Encouraged, shortly before her untimely death, by my dear friend the physician and playwright Laura Kuckes, I developed the idea of cosmological natural selection, which I published in 1992 and described in my first book, *The Life of the Cosmos.*

While I was finishing that book, another friend, Drucilla Cornell, told me to read the Brazilian philosopher Roberto Mangabeira Unger, who in a book on social theory had also argued for the evolution of laws in cosmology. She put us in touch, and after a thrilling conversation in his office at Harvard, he proposed that we collaborate on a rigorous academic book about the implications of the reality of time and the evolution of laws. That project, under way for the last five years, has been the main impetus and vehicle for the development of the ideas in this book. Thanks mainly to Roberto's clear and provocative thinking, I eventually came to appreciate the radicality of the proposal that time is real. The crux of the Epilogue — that the openness of the future should prompt human beings to reach for novel solutions to problems on every scale — was inspired largely by his writings. Those readers wanting a more rigorous formulation of the arguments sketched here are encouraged to turn to the forthcoming book, tentatively titled *The Singular Universe and the Reality of Time.*

Also in 1986, I began work on the quantization of the new formulation of general relativity invented the previous year by Abhay Ashtekar. This led to the discovery of loop quantum gravity, which I realized in joint work with Carlo Rovelli. Our technical work was motivated and framed by constant discussions on the nature of time with Abhay and Carlo along with Louis Crane, Ted Jacobson, Chris Isham, Laurent Freidel, João Magueijo, Fotini Markopoulou, Giovanni Amelino-Camelia, Jerzy Kowalski-Glikman, and Renate Loll, among many others. Indeed, I came later than several of my friends to the idea that the relativity of simultaneity would have to be given up on a cosmological scale. Antony Valentini understood many years ago that this was the cost of adopting a hidden-variables theory, and João Magueijo was already being provocative about violating relativity when we met in London in 1999. Fotini Markopoulou was the first to emphasize the importance of the inverse problem and to forcefully advocate an approach to quantum gravity in which time is fundamental and space emerges, and the main ideas of Chapter 15, including disordered locality and geometrogenesis, are due to her.

Although I'm a physicist, I have been fortunate to have been a frequent guest in the house of the philosophy of science, where I've made many friends who, over the years, have listened patiently to and critically read my attempts to think clearly about the nature of time. These include Simon Saunders, Steve Weinstein, Harvey Brown, Patricia Marino, Jim Brown, Jenan Ismael, Cheryl Misak, Ian Hacking, Joseph Berkovitz and Jeremy Butterfield, as well as my original teacher in the philosophy of physics, Abner Shimony. Julian Barbour, Jim Brown, Drucilla Cornell, Jenan Ismael, Roberto Mangabeira Unger, and Simon Saunders were kind enough to read the whole draft and give me crucial feedback.

I am grateful to Sean Carroll, Matt Johnson, Paul Steinhardt, Neil Turok, and Alex Vilenkin, whose conversations and comments on draft chapters straightened me out on cosmological issues. My work on quantum foundations described here has benefited greatly from interactions with the foundations community at Perimeter Institute, particularly Chris Fuchs, Lucien Hardy, Adrian Kent, Markus Müller, Rob Spekkens, and Antony Valentini.

Crucial encouragement and critical insight on early drafts, and much besides, was, as always, given by Saint Clair Cemin, Jaron Lanier, and Donna Moylan.

The Epilogue's concern with climate change was inspired and informed by a seminar on threshold behavior in political science, organized jointly with Thomas Homer-Dixon, of the Balsillie School of International Affairs. I am grateful to Tad and other participants, particularly Manjana Milkoreit and Tatiana Barlyaeva, for discussions and collaborations on this issue.

For an education in economics, reflected also in the Epilogue, I thank those who worked with me to organize a conference on the Economic Crisis and its Implications for the Science of Economics, held at Perimeter Institute in May 2009; and others I met at the conference and its aftermath, particularly Brian Arthur, Mike Brown, Emanuel Derman, Doyne Farmer, Richard Freeman, Pia Malaney, Nassim Taleb and Eric Weinstein.

For their friendship, collaboration, and views on self-organization, I owe an enormous debt of thanks to Stu Kauffman and Per Bak.

I am forever grateful to Howard Burton and Mike Lazaridis for the honor and unique opportunity of helping to establish Perimeter Institute for Theoretical Physics, and I am grateful also to Neil Turok for his continual encouragement of our efforts at making breakthroughs and discoveries. Every scientist and scholar should have the good fortune I have had to find an intellectual home as stimulating, diverse, and supportive of big ambitions as Perimeter Institute.

My work in physics has been generously funded by the NSF, NSERC, the Jesse Philips Foundation, Fqxi and the Templeton Foundation, to all of whom I am very thankful for the opportunity to do my work as well as to support and mentor promising young scientists.

This book is much better than it might have been thanks to comments from early readers, who gave me feedback on all or part of the manuscript. In addition to those mentioned above, these include Jan Ambjørn, Brian Arthur, Krista Blake, Howard Burton, Marina Cortes, Emanuel Derman, Michael Duschenes, Laurent Freidel, James George, Dina Graser, Thomas Homer-Dixon, Sabine Hossenfelder, Tim Koslowski, Renate Loll, Fotini Markopoulou, Catherine Paleczny, Nathalie Quagliotto, Henry Reich, Carlo Rovelli, Pauline Smolin, Michael Smolin, David Topper, Rita Tourkova, Antony Valentini, Natasha Waxman, and Ric Young.

I'm one of those writers who loves being edited—because he is painfully aware that he benefits from it. Due to the vicissitudes of publishing, no book of mine has had a more dedicated set of editors. The book owes its conception to the efforts of Amanda Cook, now at Crown Publishing, to whom I am indebted for her belief in the project as well as her conviction that it deserved the time to be sharpened and focused. Courtney Young, of Houghton Mifflin Harcourt, and Sara Lippincott, with their many comments and suggestions, have been the best editors an author could want. The book has also greatly benefited from the wise insights of Louise Dennys, of Knopf Canada. Thomas Penn's encouragement at crucial moments is also well appreciated. I am also very grateful to Henry Reich, who made many of the figures. As with all my books, I owe a huge debt of gratitude to John Brockman, Katinka Matson, and Max Brockman; without their faith in it, this book would never have come to be.

My thanks to Rodila Gregorio for continual lessons in patience, grace, and loving responsibility — and to Kai, from whom I learned everything I know about time that is not discussed here. Thanks to Pauline, Mike, and Lorna for their love and their confidence in me. Finally, words are insufficient to express my thanks to Dina, who with infinite love and patience held me together through the pressures of trying to finish a book on time on time.

INDEX

LEE SMOLIN is a theoretical physicist who has made influential contributions to the search for a unification of physics. He is a founding faculty member of the Perimeter Institute for Theoretical Physics in Waterloo, Ontario, and a graduate faculty member in the philosophy department of the University of Toronto. His previous books include *The Trouble with Physics*, *The Life of the Cosmos*, and *Three Roads to Quantum Gravity*. He lives in Toronto.